Analytical Spectroscopy Library — Volume 1

NMR for Liquid Fossil Fuels

Analytical Spectroscopy Library

A Series of Books Devoted to the Application of Spectroscopic Techniques to
Chemical Analysis

Volume 1 **NMR for Liquid Fossil Fuels**
          by L. Petrakis and D. Allen

Analytical Spectroscopy Library — Volume 1

NMR for Liquid Fossil Fuels

Leonidas Petrakis

Senior Scientist, Chevron Research Company, P.O. Box 1627, Richmond, CA 94802-0627, U.S.A.

and

David Allen

Assistant Professor, Department of Chemical Engineering, University of California, Los Angeles, Los Angeles, CA 90024, U.S.A.

ELSEVIER

Amsterdam — Oxford — New York — Tokyo   1987

ELSEVIER SCIENCE PUBLISHERS B.V.
Sara Burgerhartstraat 25
P.O. Box 211, 1000 AE Amsterdam, The Netherlands

Distributors for the United States and Canada:

ELSEVIER SCIENCE PUBLISHING COMPANY INC.
52, Vanderbilt Avenue
New York, NY 10017, U.S.A.

Chem

Library of Congress Cataloging-in-Publication Data

```
Petrakis, Leonidas, 1935-
   NMR for liquid fossil fuels.

   (Analytical spectroscopy library ; v. 1)
   Includes bibliographies and index.
   1. Liquid fuels--Analysis.  2. Nuclear magnetic
resonance spectroscopy.  I. Allen, David, 1958-
II. Title.  III. Series.
TP343.P418  1986      662'.6       86-19726
ISBN 0-444-42694-9 (Elsevier Science)
```

ISBN 0-444-42694-9 (Vol. 1)
ISBN 0-444-42695-7 (Series)

Printed in The Netherlands

4/10/2006
MMC

TP
343
P418
1987
CHEM

v

TABLE OF CONTENTS

PART TWO: CHARACTERIZING LIQUID FUELS USING NMR

PREFACE

   High resolution nuclear magnetic resonance (NMR) of liquid fuels has provided valuable information on the molecular structures present in these fuels.
The chemical insight gained through NMR studies has the potential to significantly enhance the development of processes for the utilization of fossil
energy. For this potential to be fully realized, the users of NMR information
must be able to effectively communicate with NMR experts. Conversely, the NMR
experts must understand the type of information that users will attempt to
derive from their spectra. The goal of this book is to strengthen the lines
of communication between NMR experts and users in the area of NMR of liquid
fuels.

   The book is divided into two parts. The first part presents elements of
relevant NMR phenomenology, including a definition of the most important NMR
parameters, an introduction to Fourier transform NMR and a discussion of newer
pulse techniques. This discussion of NMR phenomenology is not exhaustive, and
it is not aimed at NMR experts. Rather, it attempts to introduce sufficient
background material for the non-expert user so that NMR experts and users can
work together more efficiently. This first part concludes with many examples
from the authors' own work as well as from the literature. The review is not
intended as a complete and critical review of the voluminous literature on the
subject. Rather, it is intended to illustrate both the techniques presented
in the book and also the range of applications and the great potential of the
techniques.

   The second part of the book addresses the interpretation of NMR spectra and
is based to a very large extent on the work of the authors, who have used NMR
in a variety of applications to fossil fuels. First, an overview of data
interpretation methods is presented. Then, detailed presentations are made on
the three most common methods of interpreting NMR spectra: calculation of
average molecular parameters, average molecule construction and functional
group analysis. Another section is devoted to the use of the NMR characterizations in engineering calculations. Throughout Part Two, examples are drawn
from heavy petroleum crudes, shale oils, coal liquids and synthetic mixtures.
Comparisons and contrasts among these various fuels are made, and the potential of the NMR techniques is discussed vis-a-vis their ability to produce
data relevant to kinetics and processes for conversion and upgrading of fuels.
A section is also included on the nature of asphaltenes and contributions made
by NMR for their characterization.

   We hope that the book will appeal to a wide range of professionals. Those
who use NMR to characterize liquid fossil fuels or those who provide NMR

x

assistance to fossil fuel scientists and technologists should find it of primary interest. In addition, people interested in getting into the field, graduate students and those who manage and fund research in the utilization of liquid fossil fuels should find it of interest. One of us (LP) used some of this material in a pilot graduate course at the University of Wyoming, while the other of the authors (DA) uses the book as one of the texts for a course at UCLA on Molecular Spectroscopy of Complex Systems.

We acknowledge the cooperation and considerable help of the people who made this book possible. First, we thank Phyllis Gilbert for her great skill and patience in laying out and typing the manuscript. We also thank Carol Hicks for her preparation of the figures; Dimitris Liguras for help in assembling the final copy; Don Young for obtaining the spectra of octylbenzene; J. Kiebert and J. Friederich for their encouragement and editing of the manuscript; and Lina Petrakis for her encouragement and understanding.

Finally, we acknowledge with sincere thanks the permission granted by the American Chemical Society (Analytical Chemistry, Industrial and Engineering Chemistry Process Design and Development) and Butterworths Publishers (Fuel); Elsevier (Fuel Processing Technology) and Macmillan Journals (Nature) for allowing us to reproduce some figures that appear in the text; and the following individuals for allowing us to use material from their publications: Dr. J. Shoolery, Dr. R. Gerhards, Dr. C. Snape, Dr. W. Brey and Dr. J. Speight.

Hiller Highlands, Oakland, CA　　　Leonidas Petrakis
Los Angeles, CA　　　David Allen
May 1986

PART ONE:

ELEMENTS OF RELEVANT NMR PHENOMENOLOGY

Chapter I

THE NMR EXPERIMENT AND THE NMR PARAMETERS

In this chapter we introduce briefly the phenomenology of the nuclear magnetic resonance (NMR) experiment. The treatment will not be rigorous or
exhaustive, but rather will simply define and summarize certain concepts that
are assumed in the presentations of original research.

I.A INTRODUCTION

Conceptually, the NMR experiment is very simple even though both the classical and quantum mechanical descriptions can be quite elegant in their
rigorous presentation [Abragam, 1961; Carrington, 1967; Pople, 1959; Pratt,
1984]. We utilize both descriptions of magnetic resonance, because each provides unique insight into the nature of the experiment. Classically, the NMR
experiment may be viewed as the monitoring of reorientations of a macroscopic
magnetic dipole moment. The macroscopic moment is the resultant of the magnetic dipole moments of individual nuclei in a strong external magnetic field
H_o. According to classical mechanics, when a macroscopic magnetic dipole
moment $\mathbf{M}$ is placed in a strong magnetic field $\mathbf{H}_o$, the magnetic moment is subjected to a torque $\mathbf{T}$ which tends to align $\mathbf{M}$ with $\mathbf{H}_o$ (this notation is defined
in Table I-1).

$$\text{Torque} = \mathbf{M} \times \mathbf{H}_o \tag{1}$$

Torque is actually defined as the time derivative of angular momentum

$$\mathbf{T} = \frac{d\mathbf{p}}{dt} \tag{2}$$

The angular momentum and the magnetic dipole moment are colinear vectors, and
their magnitudes are related by the magnetogyric ratio γ

$$\mathbf{M} = \gamma\mathbf{p} \tag{3}$$

Therefore equation 1 can be rewritten as

$$\frac{d\mathbf{M}}{dt} = \gamma(\mathbf{M} \times \mathbf{H}_o) \tag{4}$$

This is the basic equation for the classical description of the NMR experiment, for it tells us the behavior of the magnetic vector in a magnetic field.
The solution of this equation, the details of which will not interest us here,
is best achieved in a rotating frame of reference and it leads to the simple

4

Table I-1

SYMBOLS USED IN THE EQUATIONS

I = spin quantum number

m = magnetic quantum number

H_o = external magnetic field

γ = magnetogyric ratio

M = magnetization vector of the ensemble of nuclei (magnetic moment vector)

E = energy of interaction between nuclear magnetic moment and external field H_o

μ = nuclear magnetic moment vector

T = torque aligning the magnetic vector with the external field H_o

p = angular momentum

t = time

H_1 = transition-inducing magnetic field

$\hbar$ = Planck's constant

σ = shielding tensor

ω = angular velocity of precession for the magnetization vector

δ = chemical shift (in ppm)

result that the magnetic moment **M** performs a precession about the applied field H_o as shown in Figure I-1. This is a well known result in classical mechanics, the Larmor theorem, and it predicts that the angular velocity of the rotation of the magnetic moment (ω_o) is proportional to the strength of the aligning field H_o, the proportionality constant being the magnetogyric ratio of the system.

$$\omega_o = \gamma H_o \tag{5}$$

There are two differences between the behavior of the macroscopic magnetic dipole moment **M** of the ensemble of nuclei and a truly classical bar magnet. The latter would rotate after being subjected to the torque, and it would continue to precess until it came to rest in a position parallel to H_o. However, the magnetic moment **M**, since it is a statistical composite of individual magnetic dipole moments, never comes to rest and continues to precess about H_o. The other difference between **M** and a classical magnet is that the classical magnet can assume a continuous range of orientations relative to H_o depending on the strength of the field. However, for the nuclear magnetic moments, the orientations are quantized, and only the 2I+1 projections on H_o are allowed, where I is the spin quantum number.

To complete the classical description of the NMR experiment, it is necessary to introduce a second magnetic field H_1, which (a) is much smaller than H_o, (b) is perpendicular to H_o and (c) rotates about H_o in the same sense as **M** with angular velocity ω_1. Of course, **M** is subject then to an additional torque **M**$\times H_1$. If the Larmor frequency of the rotating spin system, ω_o, is the same as the velocity of rotation of the field H_1, then the overall motion of **M** is a complicated one, for it is being determined by two torques, due to the two fields H_o and H_1. If the two angular velocities do not coincide, there is only instantaneous perturbation as the two meet in a given rotation, but if the two angular frequencies remain long enough in phase, then the perturbation due to H_1 can be significant. In fact, in the latter situation, energy is absorbed by the system from H_1 during the tipping of **M** from its equilibrium position, and, by monitoring H_1, NMR absorptions are detected.

An alternative description of the NMR experiment is the quantum mechanical one. In this approach, the external magnetic field H_o causes the nuclei to be distributed among the nuclear Zeeman levels according to the Boltzmann law. The applied field H_1 induces transitions between the various energy levels.

Each nucleus of the ensemble of nuclei of the system being investigated has 2I+1 energy levels, each characterized by a magnetic quantum number m

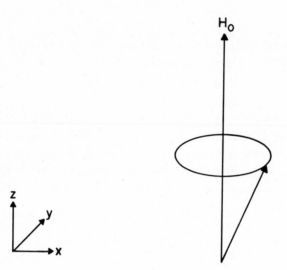

Figure 1-1. Precession of a magnetization vector about an applied field H$_\text{o}$.

$$m = I, I-1 \cdots - I+1, -I \tag{6}$$

Thus for nuclei such as proton or ^{13}C with $I=1/2$ there are only two nuclear Zeeman levels, $m = -1/2$, with the spin against $\mathbf{H}_o$ and therefore of higher energy, and $m = 1/2$, with the spin aligned with $\mathbf{H}_o$ and therefore of lower energy. In the absence of an external magnetic field the nuclear Zeeman levels are degenerate. However, in the presence of the field $\mathbf{H}_o$ the degeneracy is removed and the levels are separated, the actual separation being determined by the field strength. Each nuclear moment interacts with the field $\mathbf{H}_o$, the energy of the interaction being

$$E = -\boldsymbol{\mu}.\mathbf{H}_o = -\mu_z \mathbf{H}_o \tag{7}$$

where μ_z is the component of the nuclear moment in the z direction.
For a system with $I = 1/2$ the energy of the spin in each of the two nuclear Zeeman levels is illustrated in Fig. I-2. The two nuclear Zeeman levels are separated by

$$\Delta E = 2\mu_z \mathbf{H}_o \tag{8}$$

If the ensemble of nuclear moments, distributed as they are among the nuclear Zeeman levels, is subjected to an electromagnetic field of energy $E = \hbar\omega$, where E is equal to the energy difference between the nuclear Zeeman levels, then transitions will be induced by the field because

$$2\mu_z \mathbf{H}_o = \hbar\omega \tag{9}$$

and since $\omega = \gamma\mathbf{H}_o$

$$\omega = \frac{\mu\mathbf{H}_o}{\hbar(1/2)} = \gamma\mathbf{H}_o \tag{10}$$

The number of transitions between the nuclear Zeeman levels determines the intensity of the NMR signals. Actually, there are three kinds of transitions. The first is the spontaneous emission of radiation. The probability of spontaneous emission is vanishingly small at MHz frequencies, i.e., the typical NMR frequencies. In addition, there are the $\mathbf{H}_1$ stimulated transitions from levels $E_{m=+1/2} \rightarrow E_{m=-1/2}$ (stimulated absorption of radiation) and $E_{m=-1/2} \rightarrow E_{m=1/2}$ (stimulated emission of radiation). The probabilities for these transitions are equal to each other, and much greater than the probability for spontaneous emission. However, since the lower nuclear Zeeman level is occupied by a greater number of nuclei according to the Boltzmann distribution, there is a greater number of absorption transitions and, therefore, a net

8

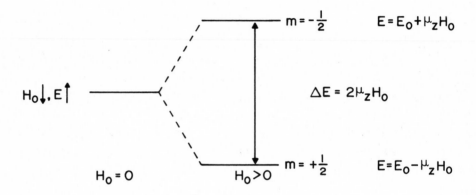

Figure I-2. Energy level diagram for spins I = 1/2.

absorption of energy from the electromagnetic field H_1. As a result, the NMR signal intensity is a direct, quantitative measure of the number of nuclei present. At field strengths and temperatures normally used in the NMR experiment, the excess fractional population of the lower energy level is only a few nuclei per million and, therefore, conventional NMR is an inherently insensitive technique. High field strengths and very low temperatures enhance the sensitivity of the technique since both high field and low temperature increase the fraction of the nuclei in the $m = 1/2$ state. Pulse Fourier transform techniques can greatly enhance the sensitivity of NMR.

Many nuclei that are important in the molecular structure of fossil fuels can be readily observed by NMR. The properties of these nuclei are listed in Table I-2.

I.B   NMR SPECTRAL PARAMETERS

Both the classical and the quantum mechanical descriptions of the NMR experiment lead to the same result, namely, that the frequency at which a given nucleus undergoes resonance depends on the strength of the field and the magnetogyric ratio γ, as shown in equation 10. If this equation was strictly valid there would be no basis for using nuclear dipole moments as probes of chemical structure, for at a given field one would simply observe infinitely narrow absorptions or lines. One line would result for each type of nucleus, and the lines would be far removed from each other. Yet, we know that the NMR absorptions are not only of finite width, but also they show a number of other features which can be sources of considerable physicochemical information. In this section we summarize the main NMR spectral parameters which relate the characteristics of NMR spectra to chemical structure.

I.B.1   Absorption Line Shapes and Line Widths

The observed NMR "line" or signal is not a Dirac delta function, but rather a signal of finite width and definite shape. Individual lines are normally either Lorentzian or Gaussian in shape. The former are encountered with gaseous or liquid systems while the signals of rigid randomly distributed solids are Gaussian. The line width and shape are determined by, among others, the following factors: magnetic field inhomogeneity; the natural line width according to the Heisenberg Uncertainty Principle; magnetic fields from neighboring magnetic nuclei; dynamical processes in which the nuclei might be involved, such as chemical exchange; and other interactions of the nuclei, e.g. the nuclear electric quadrupole moments with the molecular electric field gradient. These effects are also reflected in the relaxation times T_1 and T_2, which will be discussed below. Typically, high resolution proton NMR spectra

Table I-2

NMR PROPERTIES OF SELECTED ISOTOPES IMPORTANT IN NMR OF FOSSIL FUELS

Isotope	Natural Abundance	I Nuclear Spin, $\hbar$	μ Magnetic Moment (μ_o)	NMR Frequency in 23,487 Gauss Field, MHz	Relative Sensitivity at Constant Field	
					For Equal Numbers of Spins	At Natural Abundance
^{1}H	99.98	1/2	2.79268	100.00	1.00	1.00
^{13}C	1.11	1/2	0.70220	25.14	1.59×10^{-2}	1.76×10^{-4}
^{14}N	99.64	1	0.40358	7.22	1.01×10^{-3}	1.01×10^{-3}
^{15}N	0.36	1/2	-0.28304	10.13	1.04×10^{-3}	3.74×10^{-6}
^{17}O	3.7×0^{-2}	5/2	-1.8930	13.56	2.91×10^{-1}	1.08×10^{-5}
^{33}S	0.74	3/2	0.64274	7.67	2.26×10^{-2}	1.67×10^{-5}
^{51}V	≈100	7/2	5.1392	26.29	0.38	0.38
^{57}Fe	2.24	1/2	0.0903	3.43	3.38×10^{-5}	7.57×10^{-7}
^{61}Ni	1.25	3/2	0.746	8.90	3.53×10^{-3}	4.41×10^{-5}

may be ≈0.1 Hz wide at their half height. Broad   line   spectra · maybe   up   to
several thousand Hz wide.

I.B.2 Chemical Shifts

The precise frequency at which NMR resonance occurs for   a   given   nucleus
depends not only on the applied magnetic field but also on the chemical
environment in which the nucleus finds itself. This important   discovery   was
made early in the development of NMR and has played a key role in the utility
of NMR as a probe of molecular structure. Equation 10 that describes the   NMR
experiment refers to the bare nucleus. However, the nucleus normally finds
itself in a field of electrons. Thus, the actual magnetic field that   the
resonating nucleus experiences is contributed to by the field of the orbiting
electrons. This leads to a revised form of Equation 10, namely

$$\omega = \gamma H_{eff} = \gamma(H_o - H_{local}) = \gamma(H_o - H_o\sigma) = \gamma H_o(1-\sigma) \tag{11}$$

where σ is the shielding tensor. Actually, the tensor is determined by   a
number of factors that can be "shielding" or "deshielding". Shielding tensors
or chemical shifts have been studied theoretically and   have   been   correlated
empirically with various structural parameters.

In the years since the early detection of chemical shift, an extremely
large body of empirical and theoretical evidence has been developed relating
the observed chemical shift of a nucleus to its molecular   environment.   Normally, the chemical shifts are referred to in terms of parts per million
(ppm). A chemical shift of 1 ppm is equivalent to a σ value of 10^{-6}.

I.B.3 Indirect Spin-Spin Couplings

Early in the development of high resolution NMR it was found that often the
number of signals observed could not be accounted for by the number of chemically nonequivalent species. Furthermore, it was observed that the separation
or splitting of signals from equivalent species was independent of the applied
field. On the basis of these experiments the concept of indirect spin-spin
couplings (J-couplings) was proposed. Spin-spin couplings arise from the fact
that the magnetization state of one spin ($m = +\frac{1}{2}$ or $m = -\frac{1}{2}$) affects the magnetic field experienced by other nuclei. The fundamental understanding of
these couplings, along with an enormous number of empirical determinations of
coupling constants, made the concept of indirect spin-spin couplings of great
utility in structural elucidations, especially in providing information on the
spatial positions of nuclei.

The mechanism of indirect spin-spin couplings involves the scalar interaction of nuclear spins. It has been shown that due to symmetry considerations only the coupling between chemically equivalent nuclear spins is observable. Coupling with nuclei with $I > 1/2$ is rarely observed because such nuclei have very short relaxation times due to the interaction of their nuclear electric quadrupole moments with the electric field gradients.

The J-couplings are a great diagnostic tool, but they do complicate the appearance of NMR spectra. To what extent they complicate the spectra depends on the magnitude of the coupling constant relative to the chemical shift between the interacting groups. To clarify this point, consider the example of an ethyl group ($-CH_2-CH_3$). The chemical shift difference between the two interacting groups (i.e. the CH_2 protons and the CH_3 protons) is greater than the proton-proton coupling constant. In cases like this, the predicted coupling patterns are simple. The methyl protons in our example are split into triplet of intensity 1:2:1, while the methylene protons are split into a quartet of lines of intensity ratio 1:3:3:1. The reason for this is the following. The methylene protons are equivalent, and each can assume an orientation of $+I$ or $-I$. Since the protons are indistinguishable there are three possible orientations ($\uparrow\uparrow$; $\uparrow\downarrow$ or $\downarrow\uparrow$; $\downarrow\downarrow$). These three composite orientations perturb the methyl signal and cause it to appear as a triplet of lines of intensity 1:2:1. Similarly, the methyl protons can have four orientations ($\uparrow\uparrow\uparrow$; $\uparrow\downarrow\uparrow,\downarrow\uparrow\uparrow$,$\uparrow\uparrow\downarrow$; $\uparrow\downarrow\downarrow,\downarrow\uparrow\downarrow,\downarrow\downarrow\uparrow$;$\downarrow\downarrow\downarrow$). These four orientations perturb the methylene protons and cause them to appear as a quartet of lines of intensity 1:3:3:1. The separation between lines in each multiplet is constant (the J-constant) while the separation between the centers of the two patterns is the chemical shift difference between the two groupings of protons. For such first order NMR spectra (i.e. $\delta_A - \delta_B > J_{AB}$), a good mnemonic for the number of components and their intensities in group A is that the number of lines is $2n_B(I)+1$ where n_B is the number of interacting B spins of angular momentum I. The intensities of the components, if $I = 1/2$, are the coefficients of the binomial expansion (also given by the Pascal triangle).

The spectra of strongly coupled groups, i.e. with chemical shifts comparable to the spin-spin couplings, have complicated appearances; in order to analyze such multiplets in terms of the chemical shift and the coupling constants, one must set up the appropriate secular equation and solve it for the energies and eigenfunctions of the Zeeman levels. Selection rules determine the allowed transitions, and the corresponding intensities are proportional to the square of the matrix element of the x component of the magnetic moment. The procedures for obtaining the chemical shifts and coupling constants from such spectra are also quite well developed and computer programs exist for most common combinations that one encounters in small molecules.

I.B.4 Relaxation Times

The final two parameters that we mention here as being very useful in describing the NMR phenomena are the relaxation times T_1 and T_2. The energy absorbed by nuclei during an NMR experiment is dissipated through various modes until the nuclei return to their original energy levels. The relaxation times characterize the time scale associated with the dissipation of the energy absorbed during the NMR experiment. Both are extremely sensitive to the molecular environment and as a result they have been used extensively as probes of molecular structure and of dynamic processes.

The two relaxation times were introduced in the classical description of NMR by Bloch [1946], but they have also been discussed rigorously in quantum mechanical terms [Abragam, 1961)]. T_1 is the spin-lattice or longitudinal relaxation time. As the name indicates, in the quantum mechanical description of the experiment, spin-lattice relaxation time characterizes the time scale for the energy transfer between the spin and the "lattice". Lattice in this case means all the degrees of freedom which constitute the thermal bath into which the spins are placed and to which the spins must lose energy in order to achieve the population equilibrium among the various nuclear Zeeman energy levels. In the alternative classical description, T_1 is referred to as the longitudinal relaxation time, because it involves the transfer of energy between the ensemble of spins and the lattice as the component of the magnetization along H_o (the longitudinal component) returns towards its equilibrium value.

T_2 is the spin-spin interaction time, for it describes the interaction of two spins, without net energy transfer, that results in a loss of phase coherence as the individual spins precess about H_o. The coherent precessing of the spins is responsible for the macroscopic magnetization vector, and therefore, its decay in a plane perpendicular to H_o (where the signal is monitored experimentally) is characterized by the transverse relaxation time, as it is referred to in the classical description.

In summary then, T_1 is a measure of energy transfer while T_2 is a measure of a loss of phase coherence. Both result in a reduction of the NMR signal.

I.C ^1H AND ^{13}C NMR

The two most significant NMR probes of the structure of molecular moieties encountered in fossil fuels are ^1H and ^{13}C. Historically, high resolution ^1H NMR has proven to be the easiest and most fruitful to pursue. However, the widespread availability of Fourier transform (FT) NMR spectrometers, advances in high field magnets, computers and methodological breakthroughs, have given much impetus to the study of ^{13}C NMR. Of course, the observation of carbon NMR is most desirable, for it allows direct exploration of the structural (carbon) skeleton of the organic molecules encountered in fossil fuels.

Both ^{1}H and ^{13}C have a spin of 1/2, however, the carbon is inherently a much less sensitive nucleus than the proton. This is due primarily to the fact that only ≈1% of the carbon (^{13}C) has a magnetic dipole moment. The carbon dipole moment is also much smaller than that of the proton (0.70 vs. 2.79 Nuclear Magnetons). Table I-3 summarizes the differences between the nuclei. The sensitivity of carbon at constant field is 1.59×10^{-2} times that of protons. Thus, the exploitation of the carbon NMR had to wait for the experimental developments that allowed its easier observation. However, once the experimental difficulties of recording the carbon spectra were overcome, carbon NMR offered some significant advantages over proton NMR.

The most significant advantage of ^{13}C NMR is the much larger range of chemical shifts, 200 ppm as compared to 10 ppm for protons. The NMR of hydrogens in aliphatic environments appear over a range of 0.5 to 4.0 ppm while the corresponding carbons appear over 60 ppm (both chemical shifts from tetramethylsilane, TMS). Detailed assignments of chemical shift ranges for carbon NMR are given in Fig. I-3. Table I-4 gives a brief summary of proton chemical shifts, in a manner that is especially useful for group type analyses.

An additional advantage of ^{13}C NMR is that the carbon-proton coupling constants depend very significantly on the type of hybridization. This difference can be exploited, as we will see later, to distinguish between various types of carbon.

More detailed descriptions and compilations of chemical shifts and coupling constants can be found in Simons (1983), Farnum (1982), Breitmaier (1979), Brugel (1979) and Chamberlain (1974).

Table I-3

COMPARISON OF ^{1}H AND ^{13}C NUCLEI

	^{1}H	^{13}C
I	1/2	1/2
μ, Nuclear Magneton	2.79	0.70
Natural Abundance, %	99.98	1.11
Sensitivity at Constant Field	1.0	1.59×10^{-2}
Sensitivity at Constant Frequency	1.0	0.25
Frequency at 46.7 KG	200 MHz	50 MHz
Chemical Shift Range	10 ppm	200 ppm
	2000 Hz at 250 MHz	10^4 Hz at 50 MHz
Aliphatic	0.5-4.0 ppm	0-60 ppm
Olefinic	4.5-6.5 ppm	100-150 ppm
Aromatic	6.5-9.0 ppm	100-165 ppm
Coupling J's	0-20 Hz (H-H)	150 Hz ^{13}C-^{1}H (sp)
		160 Hz ^{13}C-^{1}H (sp^2)
		125 Hz ^{13}C-^{1}H (sp^3)

16

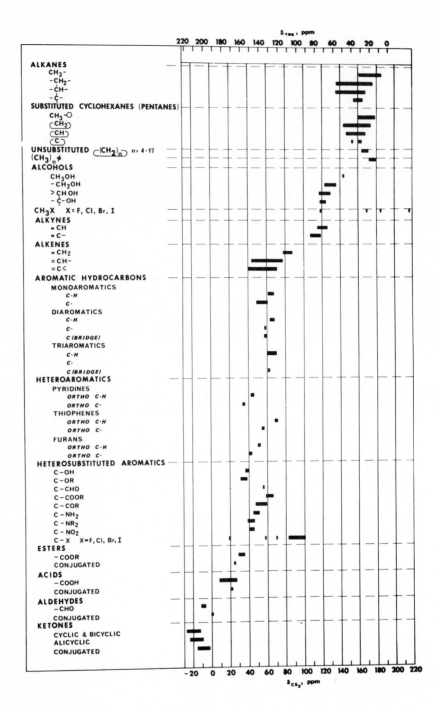

Figure 1-3. ^{13}C Chemical Shifts (Petrakis)

Table I-4

^{1}H NMR Chemical Shifts[a]

Group	^{1}H Chemical Shift[b]
CH_3-C-X[b]	0.4-2.0
CH_3-CX_n n=2,3	1.0-3.0
CH_3-Aromatic	0.4-3.4
CH_3-X	1.0-4.4
$R-CH_2-R'$ (hydrocarbon)	1.0-1.5
$R-CH_2$-Aromatic	2.0-3.5
$R-CH_2-X$	2.0-4.2
$X-CH_2-X'$	2.0-5.8
$CH_2=C-R$	4.5-5.2
$CH_2=C-X$	3.4-6.4
CH and CH_2 Cyclic	0.0-6.0
CH=C Cyclic	4.2-7.2
Aromatic	5.6-8.0
Condensed Aromatic	6.4-9.8

[a] After Chamberlain (1974)

[b] ppm from tetramethylsilane

[c] X = heteroatom

REFERENCES

Abragam, A., "The Principles of Nuclear Magnetism", Oxford University Press, London (1961).

Bloch, F., Phys. Rev., 70, 460 (1946).

Breitmaier, E., Haas, G. and Voelter, W., "Carbon-13 NMR Data", Heyden, London (1979), 3 volumes.

Brugel, W., "Handbook of NMR Spectral Parameters", Heyden, London (1979).

Carrington, A. and McLachlan, A. D., "Introduction to Magnetic Resonance", Harper and Row, New York (1967).

Chamberlain, N. F., "The Practice of NMR Spectroscopy", Plenum, New York (1974).

Farnum, A., Potts, Y. R. and Farnum, B. W., "200 MHz ^{1}H NMR Spectral Catalog of Standards Related to Low-Rank Coal-Derived Materials", DOE/FC/RI-82/4 (1982).

Petrakis, L. and Jensen, R. K., J. Mag. Res. 6, 105 (1972).

Pople, J. A., Schneider, W. G. and Bernstein, H. J., "High Resolution Nuclear Magnetic Resonance", McGraw-Hill, New York (1959).

Pratt, D. W., in "Magnetic Resonance, Introduction, Advanced Topics and Applications to Fossil Energy", L. Petrakis and J. Fraissard (Editors), D. Reidel, Dordrecht (1984).

Simons, W. W. (Editor), "The Sadtler Guide to ^{13}C NMR Spectra", Sadtler Research Laboratories, Philadelphia (1983).

Chapter II

FOURIER TRANSFORM NMR

II.A   FOURIER TRANSFORMS

Fourier transform (FT) spectroscopy, also known as interferometry or time domain spectroscopy [Farrar, 1971; Shaw, 1984; Fukushima, 1981], utilizes the well-known Fourier transform theorem to interconvert spectroscopic information between two equivalent forms. Before we delve into the precise meaning of this statement, we recall certain aspects of the Fourier transform theorem.

The description of many physical problems, such as electromagnetic waves, electric circuits, vibration, etc., is best effected through a Fourier series of the general form

$$f(x) = \frac{a_o}{2} + \sum_{n=1}^{\infty} (a_n \cos n + b_n \sin nx) \tag{1}$$

where

$$a_o = \frac{1}{\pi} \int_{-\pi}^{\pi} f(x)dx \tag{2}$$

$$a_n = \frac{1}{\pi} \int_{-\pi}^{\pi} f(x)\cos nx \, dx \tag{3}$$

$$b_n = \frac{1}{\pi} \int_{-\pi}^{\pi} f(x)\sin nx \, dx \tag{4}$$

This series is a generalized expression that allows the precise mathematical description of any periodic function. The series allows one to analyze a function, i.e., to determine which frequencies are contained in a given function; and also, it allows one to synthesize a function, i.e., to determine which frequencies must be added together to give the required function. The trigonometric expression is useful for the description of repetitive functions, but it is not particularly suitable for the description of single events.

The trigonometric form of the Fourier series can be put quite readily into the equivalent exponential form through the use of the Euler formula

$$e^{\pm inx} = \cos nx \pm i \sin nx \qquad (5)$$

Therefore

$$f(x) = \frac{a_o}{2} + \sum_{n=1}^{\infty} (a_n \cos nx + b_n \sin nx) \qquad (6)$$

becomes

$$f(x) = \sum_{n=-\infty}^{+\infty} C_n e^{inx} \qquad (7)$$

where

$$C_n = \frac{1}{2\pi} \int_{-\pi}^{\pi} f(x)e^{-inx}dx \qquad (8)$$

The exponential form of the Fourier series is more difficult to visualize, but it is simpler to use mathematically. It also leads directly to the Fourier integral, which does for a single pulse or event what the Fourier series does for a repetitive function. Thus, if the functions $\psi(x)$ and $\phi(k)$ are a Fourier transform pair, then they are related to each other through the Fourier integral in the following way

$$\phi(k) = \int_{-\infty}^{+\infty} \psi(x)e^{ikx}dx \qquad (9)$$

$$\psi(x) = \frac{1}{2\pi} \int_{-\infty}^{+\infty} \phi(k)e^{-ikx}dk \qquad (10)$$

The above equations allow the transformation back and forth between the two spaces in which a function may exist. The two forms in the two different spaces are referred to as a Fourier transform pair. A familiar example of a Fourier transform pair is a wave form in the time domain (e.g. $\cos \omega_o t$) and the spectrum in the frequency domain (a single line at ω_o in this case). In NMR spectroscopy, the Fourier integral is used to transform the NMR signal from a time domain space to a frequency domain space. In the next section we will consider how the Fourier transform mathematics are used in the interpretation of pulse NMR data. Figure II-1 shows several examples of Fourier Transform pairs.

II.B THE FOURIER TRANSFORM NMR EXPERIMENT

In order to provide a better understanding of the great power of pulsed NMR techniques, and since we will be focusing on pulse sequences in later

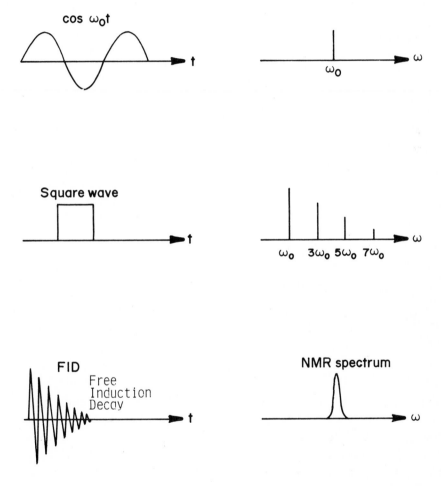

Figure II-1. Fourier transform pairs.

sections, we consider the phenomenon of Fourier Transform NMR at some detail here. As we already have shown, an ensemble of nuclei, when subjected to a magnetic field H_o will precess about the direction of the field as shown in Figure I-1. This coherent precession of the individual nuclear moments results in a macroscopic magnetization vector in the z direction, M_z. The application of another field H_1, rotating about the z direction, can cause M_z, according to the classical description of the phenomenon, to deviate from its equilibrium position. Mathematically, the situation is handled by examining this phenomenon in a rotating frame of reference, i.e., by imagining that the coordinate system shown in Figure I-1 is also rotating about the z axis at the same frequency as H_1.

When the frequencies at which M and H_1 rotate are coincident, then the macroscopic magnetization of M is tipped towards the x-y plane. In fact, the angle θ (expressed as a fraction of a complete rotation) through which M is tipped is determined by

$$\theta = \gamma H_1 t_p \qquad (11)$$

Thus, by choosing the strength of H_1 and the width of the pulse, t_p, i.e., the time during which H_1 is turned on, M can be tipped through any angle. A 90° pulse brings M in the x-y plane, while a 180° pulse causes M to become aligned in the negative z direction (Figure II-2).

The 90° pulse, which brings the magnetization vector into the x-y plane, is particularly significant because normally this is the plane in which the signal detector is situated. After the pulse is applied, a free induction signal is observed. The signal is called "free" because the applied field H_1 is off. After tipping M in the x-y plane and as soon as H_1 is turned off, the magnetization vector will relax, i.e., it will return to its equilibrium position in the z direction. The signal detected will diminish as M moves from the x-y plane. This corresponds to the Free Induction Decay signal (FID) which is the NMR signal in the time domain. The Fourier Transform mathematics discussed in Section II-A are then employed to convert this time domain signal into a frequency domain spectrum. This sequence of events is shown in Figure II-3.

In a homogeneous field the free induction signal decays exponentially with a time constant T_2, the spin-spin relaxation time, while the component of M in the z direction is restored to equilibrium with a time constant T_1, the spin-lattice relaxation time. Through a combination of pulse sequences these characteristic time constants can be measured readily. The measurement methods will not be discussed here, although both T_1 and T_2 are potentially

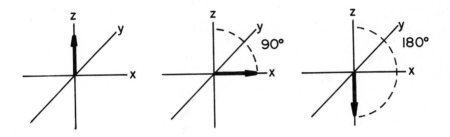

Figure II-2. 90° pulse and 180° pulse.

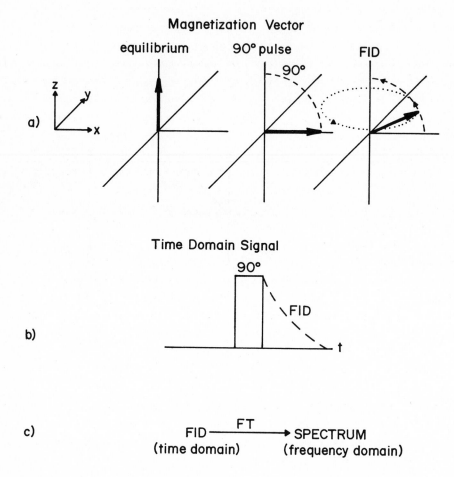

Figure II-3. The Fourier-Transform NMR Experiment

valuable handles for the characterization of petroleum and petroleum fractions.

Let us consider in greater detail the cases of proton and carbon nuclei.
Specifically we pose the question: when H_o and H_1 are applied for a given
time t what exactly happens to M_z's due to proton and carbon nuclei?

For protons, a typical H_o value for a modern instrument (200 MHz) is
46.7×10^3 gauss, and H_1 is about one thousand times smaller (46.7 gauss). Now,
$\gamma = \omega/H_o = (200 \times 10^6 \text{ Hz})/(46.7 \times 10^3 \text{ gauss}) = 4.3 \times 10^3$ Hz/gauss. In t = 1.24×10^{-6}
seconds, while H_1 is turned on, the magnetization vector will precess 250
times about the z direction. During the same time, due to the torque being
exerted by H_1 (which is in the x-y plane), the magnetization vector of the
proton will have undergone 1/4 of a rotation about the y-axis. That is to
say, it will have been tipped $\pi/2$ degrees into the x-y plane ($\gamma H_1 t_p = \theta = 0.25$
= $(4.3 \times 10^3)(46.7)(1.24 \times 10^{-6})$). This is a 90° pulse for the protons. If the
H_1 is left on for 2.48×10^{-6} sec, the M_z of the proton during this time will
have undergone 500 rotations about z, and a π rotation about x, i.e., M_z will
be in the -z direction. This is then a 180° pulse for the protons.

A similar situation exists for the carbon nuclei. Assuming that H_o has the
same value (i.e. the sample is placed in the same container) in 5 microseconds
the M_z of the carbon will have undergone 252 rotations about the z direction.
Using another H_1 in the x-y plane, in the frequency of the ^{13}C nucleus and
with a field strength of 46.7 gauss, during 5 microseconds the M_z of the carbon nuclei will have been tipped through an angle of $\pi/2$, i.e. we will have
applied a 90° pulse. For a 180° pulse this H_1 would have to be turned on for
10 microseconds. Table II-1 summarizes these experiments.

Thus, the magnetization vectors of the protons and of the carbon nuclei can
be manipulated readily. The same H_o aligns all nuclear moments, but the two
different H_1's separately can produce 90° pulses for the proton and carbon
magnetization vectors. The FID of either magnetization can be recorded individually and then transformed through an FT algorithm to yield the proton or
the carbon spectrum. It is common to obtain many FID's, after appropriate
delay periods, by applying successive 90° pulses. The FID's are then added
prior to transformation to enhance the signal-to-noise ratio. Since the 90°
pulses last for a few microseconds and the delay times are of the order of
seconds, a great number of FID's can be recorded in a brief period.

Fourier transform NMR indeed relies on a large number of pulses for its
sensitivity. However, since between pulses the nuclei may relax at different
rates, care must be taken to allow sufficient delay times. This is but one of
the complications encountered in attempting quantitative work, especially
where there are many different structural moieties of the same nucleus. The

Table II-1. $90°$ and $180°$ pulses for ^{1}H and ^{13}C Nuclei

	^{1}H	^{13}C
H_o, gauss	46.7×10^3	46.7×10^3
ω_o, Hz	200×10^6	50×10^6
γ, Hz/gauss	4.3×10^3	1.08×10^3
H_1, gauss	46.7	46.7
$90°$ pulse, sec	1.24×10^{-6}	5×10^{-6}
$180°$ pulse, sec	2.48×10^{-6}	10×10^{-6}

subject of quantitative NMR has been addressed by many specialists in the area, but particularly thorough discussions of both the theoretical and experimental details have been given by Delpuech [1984] and Martin [1980].

II.C ADVANTAGES OF FOURIER TRANSFORM NMR

In assessing the advantages of FT-NMR, we will compare it to the continuous wave spectroscopy. The main limitation of conventional CW (Continuous Wave) NMR had been the fact that it is an inherently insensitive technique. This was a particularly serious limitation with nuclei such as ^{13}C which have both a small magnetogyric ratio and low natural abundance. Signal averaging techniques allow a significant increase in the signal-to-noise ratio, but the practical limit of signal averaging had proven to be 20-30 hours. Beyond that limit gains were marginal. Any further gains involved unacceptable amounts of time, since improvement depends on the square root of the number of spectra collected, and the stability of the homogeneity of the field. Pulse, Fourier-Transform NMR has afforded the opportunity to overcome these limitations and extend considerably the usefulness of NMR.

Unlike the CW modes of the NMR experiment, the pulse or free precession technique involves the observation of the behavior of the ensemble of nuclear spins after they are subjected to a short burst (pulse) of rf (radio frequency) power at a discrete frequency. The rf pulse is of sufficient power to instantaneously excite all the frequencies of the nuclei being observed. After the pulse, the "free induction decay" signal (FID) is collected and the familiar spectrum is obtained by taking the inverse Fourier Transform. Considerable S/N enhancement can be obtained by averaging a series of free induction decay signals. This can be accomplished over a much briefer period because in a sense the entire spectrum is excited simultaneously, and the time between successive pulses is also much shorter.

Thus, the advantages of FT NMR over CW NMR are a considerable saving of time and a large increase in S/N (signal-to-noise) (Farrar and Becker). For example, if a conventional CW spectrum requires 500 seconds while the interferogram can be recorded in 0.5 second, then by storing 1000 interferograms within the 500 second span a S/N enhancement of $(1000)^{1/2}$ is effected. The ability to carry this out readily has established FT NMR as an indispensible analytical tool for fossil fuels.

References

Delpuech, J. J., in "Magnetic Resonance, Introduction, Advanced Topics and Applications to Fossil Energy", L. Petrakis and J. P. Fraissard, Editors, D. Reidel, Dordrecht (1984).

Farrar, T. C., and Becker, E. D., "Pulse and Fourier Transform NMR", Academic Press, New York (1971).

Fukushima, E., and Roeder, S. B. W., "Experimental Pulse NMR - A Nuts and Bolts Approach", Addison-Wesley, Reading, MA (1981).

Martin, M. L., Delpuech, J. J., and Martin, G. L., "Practical NMR Spectroscopy", Heyden, London (1980).

Shaw, D., "Fourier Transform NMR Spectroscopy", Elsevier, Amsterdam (1984).

Chapter III

PULSE TECHNIQUES

III.A INTRODUCTION

Most fuels are complex mixtures of hundreds to thousands of components, and therefore yield NMR spectra with broad bands, as shown in Figure III-1. Because of the broad overlapping bands, the amount of structural information that can be extracted from conventional ^{1}H and ^{13}C spectra is limited. More structural information can be obtained, however, by applying some of the special pulse sequences that have been developed for ^{1}H and ^{13}C FT-NMR. In this section we begin by describing the information provided by conventional ^{13}C spectroscopy. Then we will describe how particular FT-NMR pulse sequences can be used to obtain additional information regarding the relative amounts of CH_x (x = 0,1,2,3) present in fuels. In particular, we will examine the information provided by the Partially Coupled Spin Echo (PCSE), Insensitive Nucleus Enhancement by Polarization Transfer (INEPT), and Distortionless Enhancement by Polarization Transfer (DEPT) pulse sequences. In addition, we will examine briefly two-dimensional (2-D) NMR and the information provided by this experiment.

III.B DECOUPLING AND SPECTRAL SIMPLIFICATION

The ^{13}C NMR spectra of even the simplest hydrocarbons encountered in fossil fuels are quite complex. Of course, the aromatic and aliphatic regions are separated sufficiently well that an immediate determination can be made of the relative amounts of carbon atoms in aliphatic and aromatic environments. However, the assignment of spectral features to specific molecular environments within these regions is not obvious. Consider, for example, the ^{13}C NMR spectra of pentane, hexane and heptane isomers. Gerhards (1984) has separated the CH_3, CH_2, CH and quaternary carbon (C_q) for these compounds and the chemical shifts are shown in Figure III-2. The degree of overlap between the carbon types is striking and highlights the difficulties associated with detailed spectral assignments using conventional spectra. In order to resolve the carbon types associated with even these simple mixtures, specialized pulse sequences must be employed.

The situation is much more complex with actual fossil fuels. Yet, being able to determine the relative amounts of the various CH_x's is of paramount importance for fossil fuels. The degree of branching, condensation and substitution can be deduced from CH_x determinations and these factors influence

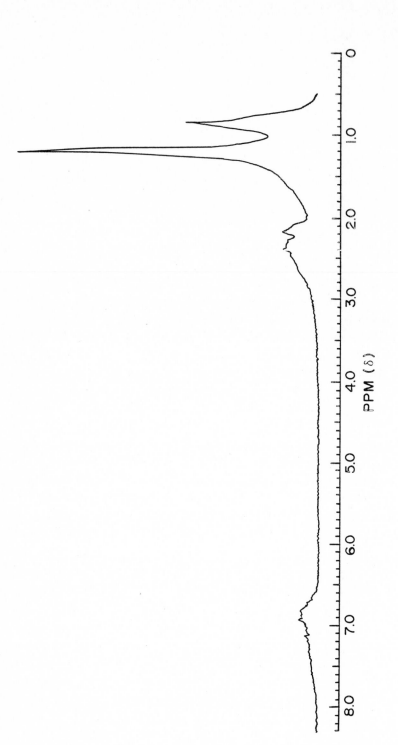

Figure III-1. ^{1}H NMR Spectrum of a Coal Derived Liquid (Note the broad bands and the limited structural information that is obtained from the spectrum.)

31

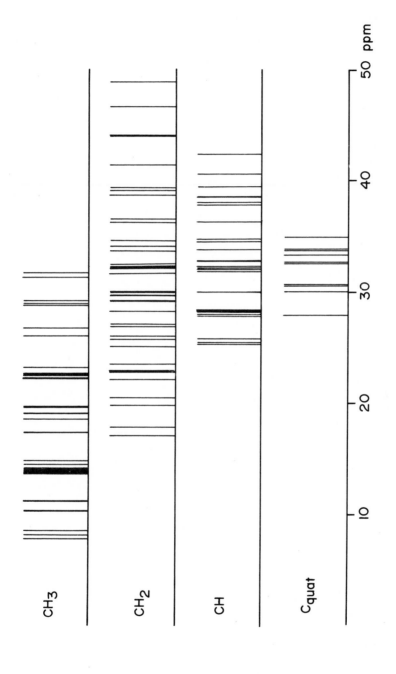

Figure III-2. ^{13}C Chemical Shifts of Carbons in Pentane, Hexane and Heptane Isomers (Gerhards)

profoundly the behavior of the mixtures. Therefore, what is needed are techniques that allow the quantitative determination of CH_x in these complex hydrocarbon mixtures. Such techniques have been developed and all of them take advantage of the interaction between the carbon nucleus and the protons which are attached to each carbon.

We discussed earlier the simple appearance of spin multiplets, which can be used to identify interacting spins. For example, the ^{13}C NMR spectrum of a methyl group will be a quartet of lines with relative intensity 1:3:3:1. This is due to the indirect carbon-proton spin-spin coupling, i.e. the ^{13}C nucleus interacting with the three equivalent protons which are attached to it. This heteronuclear J-coupling arises in an entirely analogous fashion to the homonuclear J-coupling that was discussed earlier for the protons of the ethyl group. If during the recording of this carbon quartet there is an appropriate disturbance of the carbon-proton interaction (heteronuclear decoupling), the quartet will collapse into a single peak. This decoupling of carbon and protons is accomplished by irradiating the sample over all ^{1}H NMR frequencies. In addition to simplifying the spectrum (one peak instead of four), the signal-to-noise ratio would increase because all the intensity would be concentrated into a single peak. There are additional bonuses to be gained in this decoupling experiment: the observed single peak would be enhanced by up to a factor of three due to the Nuclear Overhauser Enhancement (NOE) (Carrington, 1967); and the peak would narrow even more if there had been long range couplings. Thus the incentives to decouple the carbon spectra are high due to the simplification and signal enhancement that result. Figure III-3, (Gerhards, 1984) shows the considerable simplification and signal-to-noise enhancement that results when the proton-carbon couplings are removed using broad band decoupling. Although broad band decoupling results in some spectral simplification, what is really needed is an approach that allows the discrimination of the various CH_x groups. The following sections introduce briefly some elegant modern techniques that are now finding widespread use.

III.C SPIN ECHOES

We have shown earlier that the application of a brief, intense magnetic field H_1 perpendicular to H_o exerts a torque on the magnetization vector M and tends to realign it in the direction of the applied field H_1, away from its equilibrium position in the z-direction (Figure II-2). To what extent the magnetization vector is tipped away from its equilibrium position is determined by the strength of H_1 and the time during which it is turned on.

Figure III-4 shows the sequence of events that take advantage of this phenomenon and lead to the formation of spin echoes. Because this is impor-

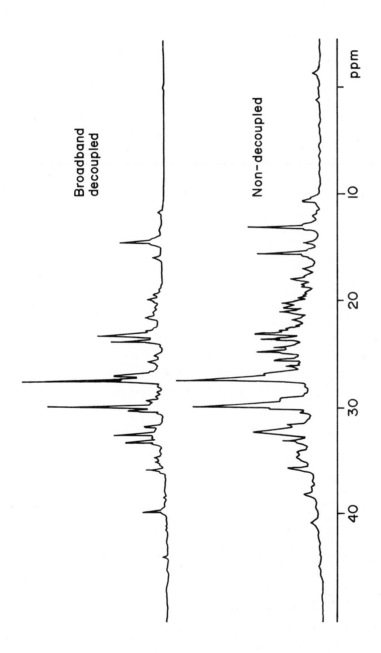

Figure III-3. Non-decoupled and Broadband Decoupled Spectra of a Liquid Fuel (Gerhards)

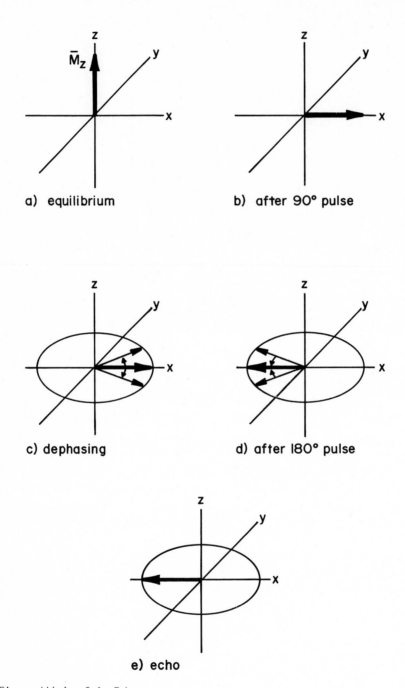

a) equilibrium   b) after 90° pulse

c) dephasing   d) after 180° pulse

e) echo

Figure III-4.   Spin Echoes

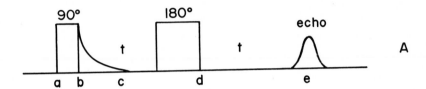

A

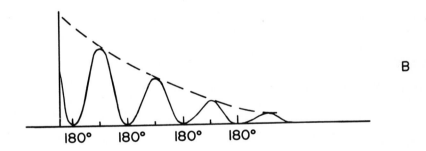

B

Figure III-5. The Decay of Spin Echoes
In A) the points a-e correspond to the magnetization states a-e
given in Figure III-4.

tant to the understanding of modern pulse sequences, we examine in detail the formation of spin echoes, first described by Hahn (1953).

The application of a 90° pulse brings the magnetization vector from its equilibrium position (Fig. III-4,a) in the +z direction to the +x direction in the xy plane (Fig. III-4,b). Because the magnetization vector **M** is the composite of a large number of individual magnetic dipoles, after the **M** is brought into the +x axis of the x-y plane, a dephasing begins due to spin-spin coupling. The reference frame and the composite vector **M** are rotating at some frequency, ω_o, but the individual magnetization vectors are rotating at frequencies slightly different than ω_o due to spin-spin coupling. This means that the individual spins are moving in the relative directions shown in Figure III-4,c, since some go faster and some slower than the composite vector **M**. If at this time there is a 180° pulse applied in the y direction, the "fan" of the dephasing spins is flipped to the -x direction (Fig. III-4,d), and since the spins continue to move in the directions that they were moving prior to the 180° pulse, now they in effect move towards each other. After a period of time, they will be coincident in the -x direction. Thus, they will form a spin echo (Fig. III-4,e).

Hahn (1953) used an elegantly simple analogy to explain this effect. Let us assume, he stated, that we have a number of runners in phase, i.e. all at the starting line of a track. When the starter's gun is fired, they run, each at his own individual speed, and after time t they are spread out, i.e. they have become "dephased". At that time the gun is fired again (the 180° pulse), and the runners are obliged to turn around and run, each with his own speed, as before. Thus, at the end of another time interval t all the runners will be at the starting line together. Their formation at this point may be considered an "echo" of the starting situation.

Getting back to the spin echo shown in Figure III-4, if another 180° pulse is applied after the first echo has been formed and has started dephasing, another echo will be formed in the +x direction. This echo will be diminished in intensity compared to the first echo. This process can be continued until the signal has decayed completely. The constant for the decay of the echoes is a measure of T_2, the spin-spin relaxation time (Fig. III-5).

III.D PARTIALLY COUPLED SPIN ECHO (PCSE) SPECTRA

The Partially Coupled Spin Echo (PCSE) sequence is the simplest of a large number of pulse sequences developed with the goal of simplifying spectra and thereby increasing the information value of NMR spectra. PCSE (sometimes also known as APT, from Attached Proton Test, among NMR practitioners) has proven particularly useful in making quantitative determination of CH_x groups. Specifically, PCSE can discriminate between (i) carbons with zero or two protons and (ii) carbons with one (methine) or three (methyl) protons. The technique takes advantage of the different behavior of the ^{13}C magnetization vector after a 90° pulse. The difference in behavior depends on whether the carbon has an odd (1,3) or even (2,0) number of protons attached to it. Because of its importance, and also because it can help in understanding the other more complex pulse sequences, we describe here in some detail the behavior of the ^{13}C and proton magnetization vectors in a PCSE pulse sequence.

Figure III-6 shows the pulse sequence in the two channels (proton and ^{13}C) used in performing the experiment. The applied magnetic field, we will assume, is 46.7 kg, which is typical of modern instruments. At that field, the ^{13}C resonance frequency is 50 MHz and the ^{1}H resonance frequency is 200 MHz, so there is a ^{1}H frequency or channel and a ^{13}C channel. The experiment begins with a 90° pulse in the ^{13}C channel. An $\mathbf{H_1}$ field of strength 46.7 gauss is applied in the ^{13}C channel for 5 µs while the ^{1}H channel remains off. At the end of the 90° pulse, the magnetization vector due to the carbon nuclei is found in the xy plane, and it begins to evolve or dephase. The evolution of the magnetization vectors will be different for the various CH$_x$'s. At precisely $\tau = 1/J$ we turn on the high powered proton channel to effect broad-band decoupling and simultaneously in the carbon channel we apply a 180° pulse. While the proton channel irradiation "scrambles" the proton energy levels and removes the proton-carbon spin-spin coupling, the carbon spins undergo an inversion (from the +x to -x) and proceed to refocus, culminating in an echo at time 1/J later. At that time the proton channel is turned off, and the FID in the carbon channel is recorded. These FID's then are Fourier-transformed to yield the carbon spectra without the splittings due to the J_{H-C}. In addition, because during the evolution period $\tau = 1/J$ the different CH$_x$'s evolved differently, the echoes due to each grouping will have a different phase. Thus, a separation of the CH$_x$'s is achieved. Precisely, why this happens is what we discuss in the following pages.

Depending on the hybridization of the CH$_x$, the J_{C-H} may vary between 125 Hz (sp^3) and 250 Hz (sp^2) (Table I-3). In fuel samples we deal primarily with aliphatic groups, and the J_{H-C} = 125 Hz. Therefore, we examine the behavior of the carbon magnetization vectors at 4 ms ($\tau = 1/2J$) after the application of the 90° pulse; at τ = 8 ms ($\tau = 1/J$); and after the refocusing 180° pulse (Fig. III-6).

Figure III-7 shows the evolution of the carbon magnetization vector due to quaternary carbons (no protons attached). At $\tau = 0$, the 90° pulse brings the magnetization vector into the xy plane in the +y direction. Since there are no protons attached to this carbon, there is no dephasing of the carbon magnetization vector due to any proton-carbon spin-spin coupling. Thus, as shown

38

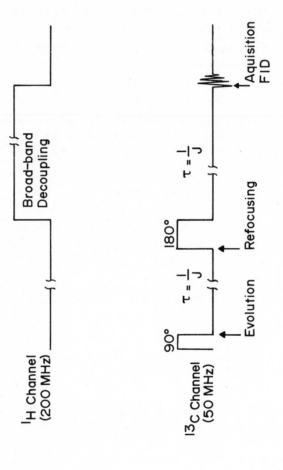

Figure III-6. The PCSE Pulse Sequence

in Fig. III-7, the vector at τ = 8 ms is in the +y direction. In the rotating frame the $C_{quaternary}$ is stationary.

Now let us examine the behavior of the carbon vector due to the CH group (Fig. III-8). This, of course, has a carbon spectrum which is a doublet due to the J_{H-C} = 125 Hz. In the non-decoupled spectrum a doublet is observed, the frequency of one line +J/2 Hz the other -J/2 Hz shifted from the center of the doublet. Thus, in the rotating frame, one component of the doublet rotates faster than the frame itself, the other more slowly. During the evolution period one frequency vector rotates clockwise and the other counterclockwise in this frame. Since both frequencies are shifted by the same amount from the center of the multiplet, the positions of the corresponding vectors are symmetric to the y-axis at any moment. As is shown in Fig. III-6 the duration of the evolution period is exactly 1/J ms. After this time interval both vectors have completed exactly one half of a circle and superimpose as at the beginning of the evolution period, but now they are pointing in the opposite direction. At this moment the broad-band decoupler is switched on (Fig. III-6) and the C-H coupling is no longer possible. The dephasing due to the spin-spin coupling of the two frequency vectors for the CH-group stops and they remain superimposed for the rest of the sequence. The C_{quat} frequency vector in the rotating frame (Fig. III-7) at this moment is exactly opposite to the superimposed vectors representing the CH-group (Fig. III-8). If a spectrum could be taken immediately following the evolution period, nonsplit signals would be observed for C_{quat}-groups as well as for CH-groups but their amplitudes would be opposite to each other.

Because there is some inhomogeneity in the static magnetic field H_o a 180° pulse in the ^{13}C-channel (Fig. III-6) is used to refocus the spins. The refocusing is best when a time interval as long as the evolution period has passed. So this refocusing period again lasts 1/J ms.

Similar considerations also apply to CH_2 and CH_3-groups: a CH_2-signal splits into a triplet if the carbon-proton spin-spin coupling is not removed (Fig. III-9). The center line has the same frequency as the rotating frame, the outer lines are J Hz apart. In the rotating frame the latter start to move clockwise and counterclockwise as discussed before. Since they are shifted J Hz from the center of the multiplet they can both finish a full circle during the evolution period 1/J. There they superimpose the center line of the triplet that remained stationary during the whole period. Thus the resulting vector is in phase with the frequency vector of the quaternary carbon signals. In the acquired spectrum CH_2-signals therefore will have the same sign as the C_{quart}-signals.

CH_3-signals (Fig. III-10) will show inverted amplitudes like CH-signals. In the CH_3-quartet the inner lines are 1/J Hz apart from each other and J/2 Hz

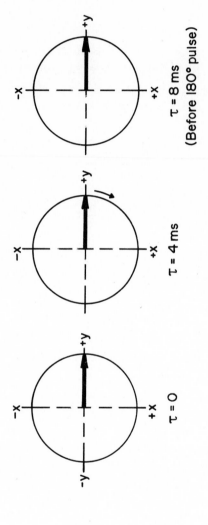

Figure III-7. Evolution of Quaternary Carbon Magnetization Vector in a PCSE Sequence

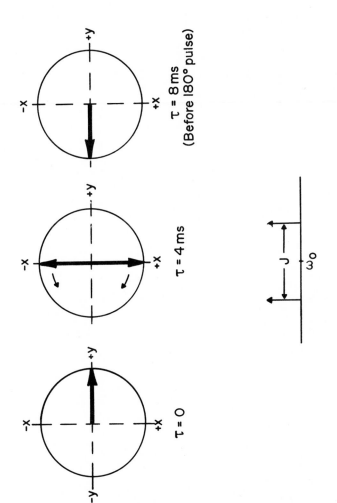

Figure III-8. Evolution of the CH Carbon Magnetization Vector in a PCSE Sequence

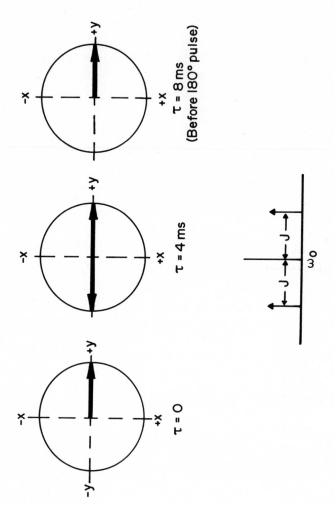

Figure III-9. Evolution of the CH$_2$ Carbon Magnetization Vector in PCSE Sequence

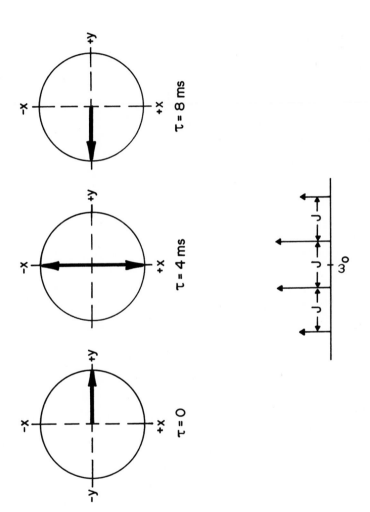

Figure III-10. Evolution of the CH$_3$ Carbon Magnetization Vector in PCSE Sequence

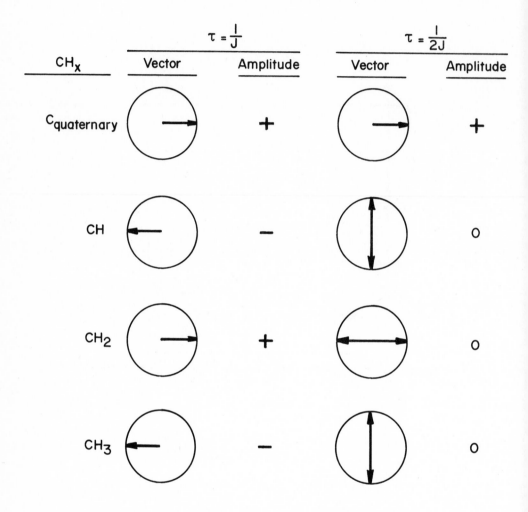

Figure III-11. The Relative Senses of CH_x Vectors in PCSE Sequence

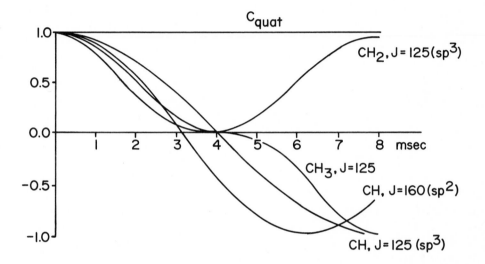

Figure III-12.   The Amplitude of CH$_x$ Vectors for Different Periods of Evolution and Different J$_{CH}$'s

from the center of the multiplet and thus they can be treated like the CH-doublet. The outer lines are separated by $3 \cdot J/2$ hertz from the center. The corresponding vectors can each rotate exactly 1.5 circles in the rotating frame during the evolution period and then they meet the vectors of the inner lines at the same position. All four vectors then superimpose. The resulting vector again is inverted to the starting orientation.

Following the same explanation it can be shown that by reducing the length of the evolution period as well as of the refocusing period by a factor of 2 to a value of $1/(2J)$, C_{quart}-groups can be observed in the acquired spectrum exclusively. For every other CH-group the corresponding frequency vectors then will exactly cancel in the rotating frame and thus prevent the appearance of these signals in the spectrum (Fig. III-11).

There is nothing magical about the values of 8 and 4 ms in terms of which we have analyzed the behavior of the magnetization vector except for the fact that they quantify the evolution period in terms of a 125 Hz spin-spin coupling constant. In fact, any time interval may be chosen for the evolution period and then the relative sense (positive, negative) of the magnetization vectors will depend on the value of the J. This is shown in Fig. III-12, where sp^3 hybridization situations are shown ($J = 125$) and sp^2 ($J = 160$). With no evolution time all carbon magnetization vectors would be at the same +x orientation in the xy plane following a 90° pulse. Regardless of the evolution time, the C_{quat} remains invariant. At $\tau = 1/2J$, CH, CH_2 and CH_3 (all sp^3 hybridization and $J_{HC} = 125$) have no resultant magnetization vector and only the quaternary carbons are recorded in a PCSE experiment. Clearly, PCSE is a powerful technique that can provide quantitative information on the various types of carbon present in a multicomponent mixture. In summary then, the PCSE method can be used to assign aliphatic peaks in ^{13}C spectra to either quaternary C, CH, CH_2 or CH_3.

III.E POLARIZATION TRANSFER TECHNIQUES

In the previous sections we described in some considerable detail the nature of the spin echoes and the partially coupled spin echoes which allow the quantitative determination of various types of carbons. Although PCSE has proved to be very valuable, additional significant contributions have been made using other pulse sequences. The carbon nucleus remains inherently a very insensitive nucleus and one would like to overcome that drawback as much as possible. Polarization transfer techniques accomplish that readily. In polarization transfer techniques, advantage is taken of the interaction of two nuclear spins (carbon and protons through the spin-spin J-coupling) and while the less sensitive but more interesting nucleus is being observed (carbon) the much greater polarization of the more sensitive nucleus (proton) is being transfered to the carbon nucleus. The benefit of doing these polarization transfer pulse sequences is actually twofold. Not only is the proton polarization transfered to the carbon, but also, the delay for the relaxation of the carbon nucleus can be shortened considerably. Thus, in a given time, many more transients of the carbon nucleus may be obtained for subsequent Fourier transformation. In effect, carbon NMR spectra are being recorded, but it is as if the carbon nucleus were as sensitive as the proton nucleus.

We will address two such polarization transfer techniques: Insensitive Nuclei Enhancement by Polarization Transfer (INEPT) and Distortionless Enhancement by Polarization Transfer (DEPT). In both cases the resulting signal-to-noise is much greater than that achieved by PCSE using the same amount of instrument time. Now there is a price to be paid for this gained advantage. Since polarization transfer is effected via the J-coupling, the quaternary carbons cannot be observed in INEPT and DEPT sequences. Also, polarization transfer sequences depend very significantly on the setting of proper interval times which, of course, are determined by J. Therefore, only those carbons whose actual J does not differ greatly from the assumed J will be observed. On the other hand, such a possibility affords the opportunity to simplify the spectra and do what is known as "spectra-editing".

III.E.1 Insensitive Nucleus Enhancement by Polarization Transfer (INEPT)

The pulse sequence for the INEPT technique is shown in Fig. III-13. Figure III-14 shows what happens to the magnetization vector of proton(s) coupled to a carbon nucleus, when a pulse sequence, as shown in Fig. III-13 is used (Gerhards). Initially the proton magnetization vector is rotated into the xy plane (+y direction) through the use of a 90° pulse in the proton channel applied along the x-direction. The carbon magnetization vector remains in its equilibrium position in the +z direction. During the evolution time $\tau = 1/4J$, the proton vector dephases due to the J_{CH}, and its two components move 1/4 of a rotation, one clockwise and the other counterclockwise, quite analogous to the PCSE experiment. The sequence now employs two simultaneous 180° pulses in the proton and carbon channels making the vectors mirror images. At the end of an additional 1/4J evolution period the proton magnetization vector components are in the +x and -x directions. The 90° pulse in the y-direction (proton channel) rotates these vectors in the +z and -z directions. The component moving to the +z direction corresponds to the normal equilibrium population with more numerous spins in the lower energy level. The component moving in the -z direction corresponds to a population inversion. (If a magnetization vector is in the -z direction, the spin population is inverted, i.e. there are more spins in the higher energy level than in the lower energy level.) This rotation of the proton vectors into the +z and -z direction

48

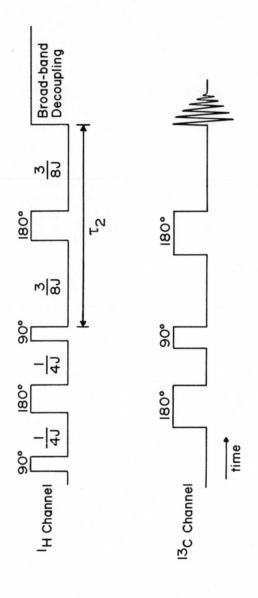

Figure III-13. Pulse Sequence for INEPT

90° ^{1}H pulse about x axis

180° ^{1}H pulse about y axis
180° ^{13}C pulse about x or y axis

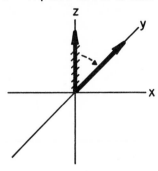

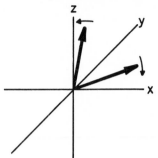

$\tau = \dfrac{1}{2J}$

$\tau = \dfrac{1}{2J}$ 90° ^{1}H pulse about the y axis

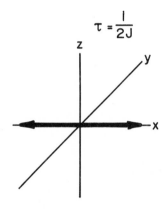

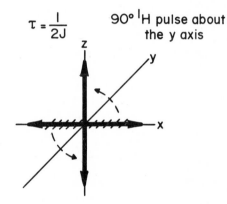

Magnetization vectors

^{1}H ^{13}C

Figure III-14. Behavior of Magnetization Vectors in an INEPT Sequence

causes the carbon spins to have population differences appropriate to the proton vectors. In other words, negative proton vectors cause negative carbon vectors since both carbons and protons share common energy levels. The simultaneous 90° pulse in the carbon channel would bring the carbon vectors (now enhanced by the proton polarization transfer) into the xy plane, where after further evolution for $\tau = 3/8J$ and refocusing with 180° pulses, they have their FID's recorded. During FID collection the proton channel is broad band decoupled to arrest any further dephasing due to J_{CH}. The net effect of the polarization transfer is the enhancement of the carbon signals by a factor of γ_H/γ_C.

There exist various versions of INEPT. If the total refocusing period is 1/4J (INEPT-1) all CH_x signals are observed in the positive direction and without splittings. If the total refocusing period is 1/2J (INEPT-2) only the CH signals are observed. If the refocusing period is 3/4J (INEPT-3) all CH_x signals will be singlets and the CH_x (x=1,3) will have positive amplitudes while CH_2 will have negative amplitude. Table III-1 summarizes these results (Gerhards, 1984).

III.E.2 Distortionless Enhancement by Polarization Transfer (DEPT)

The Distortionless Enhancement by Polarization Transfer (DEPT) technique was developed as an improvement on INEPT (Doddrell, Soerensen). The pulse sequence utilized for this experiment is shown in Fig. III-15. In this sequence the carbon atoms are excited indirectly by polarization transfer from the hydrogen atom(s) to which they are coupled. As a result, DEPT does not give signals due to quaternary carbons. On the other hand the carbon magnetization vector is greatly enhanced by polarization transfer and the carbon relaxation time bottleneck is overcome leading to great improvements in signal-to-noise for the same amount of recording time.

The initial part of the DEPT sequence is quite analogous to that of INEPT. That is, the proton and carbon magnetization vectors are manipulated and put in place to effect the polarization transfer by simultaneous 90° pulses as in the INEPT sequence. In the DEPT sequence the 90° pulse in the proton channel is replaced by a 180° pulse. Following a second evolution by $\tau = 1/2J$ there is a 180° refocusing pulse in the carbon channel, while simultaneously in the proton channel a characteristic θ-pulse is applied. This characteristic θ-pulse modulates the intensities of the CH_x signals, with the signal amplitudes being proportional to a function of θ as shown in Table III-1. Figure III-16 shows the modulation of the signals due to the various CH_x groups.

It is clear then from Table III-1 and Fig. III-16 that the DEPT sequence provides a very powerful technique for doing spectral editing and for obtaining individually the quantitative information for each carbon type in a com-

Table III-1. Summary of Intensities of CH_x Signals*

Method	Values of Parameters	Amplitude			
		CH_3	CH_2	CH	C_{quat}
PCSE-1	$\tau = 1/(2J)$	0	0	0	+a
PCSE-2	$\tau = 1/J$	+a	−a	+a	−a
INEPT-1	$\tau_1 = 1/(4J)$; $\tau_2 = 1/(4J)$	+a	+a	+a	0
INEPT-2	$\tau_1 = 1/(4J)$; $\tau_2 = 1/(2J)$	0	0	+a	0
INEPT-3	$\tau_1 = 1/(4J)$; $\tau_2 = 3/(4J)$	+a	−a	+a	0
DEPT-1	$\tau = 1/(2J)$; $\theta = 45°$	+a	+a	$+a/\sqrt{2}$	0
DEPT-2	$\tau = 1/(2J)$; $\theta = 90°$	0	0	+a	0
DEPT-3	$\tau = 1/(2J)$; $\theta = 135°$	+a	−a	$+a/\sqrt{2}$	0

*refer to Figures III-13 and III-15.

Table III-2 Spectral Editing Procedures*

C_{quat}	unobserved
CH	DEPT-2
CH_2	((DEPT-1)-(DEPT-3))/2
CH_3	$((DEPT-1)+)DEPT-3) - \frac{2}{\sqrt{2}}$ (DEPT-2))/2

*DEPT-1, DEPT-2 and DEPT-3 experiments are defined in Table III-1 and Figure III-16.

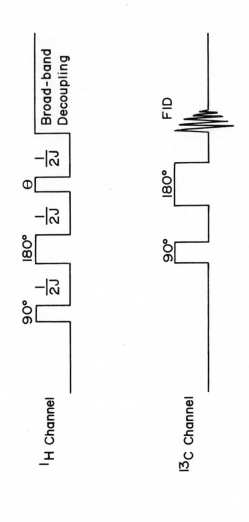

Figure III-15. Pulse Sequence In DEPT

53

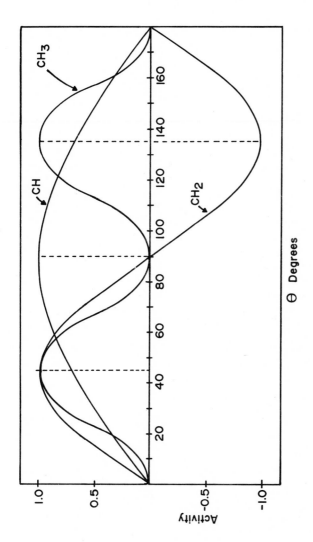

Figure III-16. Relative Amplitudes of the Magnetization of CH_x as a Function of the Width of the θ Pulse

-plex mixture. For example, if $\theta = 90°$, then the only signal observed is that due to CH, since both the CH_2 and CH_3 signals have a node at that point (Fig. III-16). On the other hand, if we measure the signal at $\theta = 135°$ and $\theta = 45°$, then the composites of the observed signals, added according to Table III-2 yield individually the CH_x signal. The only requirement is to make the amplitude measurement at three different values of θ.

III.E.3 Quantitative Results

The results of the spin echo and polarization transfer experiments described above depend directly on the knowledge of the coupling constants, J, of the carbon atoms within the sample under investigation. These constants are necessary to calculate the optimal delays in the pulse sequences. In typical samples there is a range of coupling constants present, however, a pulse sequence can principally fit only one coupling constant. Every coupling constant deviating from the value actually used will cause the corresponding signal to show an intensity other than the ideal. The frequency vectors in the rotating frame then will no longer exactly superimpose nor cancel. Vector addition then produces either shorter vectors corresponding to reduced intensities, or non-zero vectors. Fortunately, the J-values of substances in coal and petroleum derived products generally are very close together. Data available in the literature (Bremser) for pure hydrocarbons that are the main components of these products give J-values between 155 and 170 Hz for carbon atoms in aromatic bonds and J values between 120 and 130 Hz for those in aliphatic bonds. Corresponding to these data, two independent experiments generally should be performed with products made from coal or petroleum. One with the assumption J = 160 Hz giving the best results for the aromatic carbon atoms, and another with J = 125 Hz for the aliphatic carbon atoms.

However, this is necessary only if the intensities of the recorded signals are to be evaluated quantitatively or if a technique is used to suppress one or the other CH_x-signals by cancellation of the individual frequency vectors in the rotating frame.

If the experiment performed is only used to determine the nature of the individual signals, usually one single experiment is needed, assuming an average coupling constant J = 140 Hz. The actual coupling constants of aromatic and of aliphatic carbon atoms then are still close enough to differentiate between the CH_x-groups.

If quantitative results are necessary, the vector model can help to explain what effect deviations will have on the intensities of the individual signals in the recorded spectrum. As an example, consider the consequences of a 10 Hz deviation between the delay and J values to a CH_2 signal in a PCSE experiment. Assuming an actual coupling constant of 135 Hz and a value of 125 Hz used in determining the delays in the pulse sequence results in the following: instead of exactly one cycle the two frequency vectors representing the outer lines of the CH_2 triplet in the time interval $1/J = 1/125$ seconds will pass $135/125$ cycles and then will be $28.8°$ away from the starting position in the rotating frame. The vector of the center line has remained remained stationary in the rotating frame over the entire evaluation period. The total signal intensity then can be calculated by adding up all frequency vectors. Due to the symmetric distribution of the vectors with respect to the center vector, all vector components perpendicular to its direction exactly cancel. Assuming unit length for the frequency vectors representing the outer lines of the triplet, the length of their components pointing in the same direction as the frequency vector of the center line turns out to be the cosine of $28.8°$ which is 0.88. The vector sum then is 2 (for the double intensity of the center line) plus 2(0.88) (for both outer lines) resulting in 3.76. Normalizing to a standard intensity 1 that would be observed if the actual and the assumed J-values are the same, in this case the intensity is only 0.94. With the same assumptions the intensity of a CH group would be reduced to 0.97 and that of a CH_3 group to 0.91.

III.F 2-DIMENSIONAL NMR

Two-dimensional NMR was proposed by Jeener (1971), but only during the last few years has it begun to find widespread application. It promises to play a significant role in structural determinations of organic molecules. 2-D NMR (Bax, 1982,1984) is able to separate many of the interactions present in a conventional NMR spectrum (e.g. chemical shift from coupling constants), and therefore it can do what conventional NMR can, but more simply and quickly. The two dimensions of the 2-D NMR are: a) the normal spectrum, and b) a perturbation of the normal spectrum as a function of "evolution time". Actually, in the 2-D NMR experiment one recognizes several distinct time intervals (Fig. III-17):

Preparation time: it is a waiting period for several T_1's

Evolution time: a variable evolution time, t_1, following a $90°$ pulse

Mixing time: a second $90°$ pulse that rotates the evolving magnetization into the xz plane

Detection time: time for the detection of the signal, t_2.

In a 2-D NMR experiment, the signal is a function of two times t_1 and t_2. For each t_1 (evolution time) a series of FIDs are recorded during the detection time t_2. The basic idea is to examine the effect of varying the

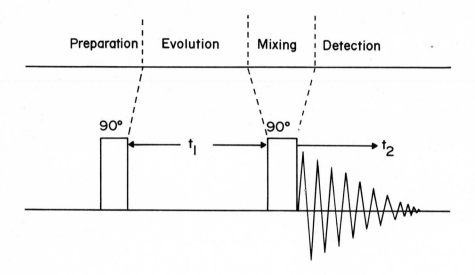

Figure III-17. Pulse Sequence for 2-D NMR

evolution period, and we have seen from our examination of 1-D pulse sequences that varying evolution times can reveal structural information such as coupling constants. By making the appropriate choices for t_1 and t_2 and the pulse sequences, the two time variables can be transformed into a number of structural variables. Thus, the two dimensions in 2-D NMR depend on the particular pulse sequences that are used. In fact any pair of interactions would be suitable for 2-D spectroscopy. However, the following pairs have proven so far to be the most useful.

a)   J-Resolved Spectra

   ¬ They separate the chemical shift δ_x from J-couplings.

   - The first dimension is the J_{H-H} (homonuclear) or J_{C-H} (heteronuclear).

   - The second dimension is the ^{1}H-decoupled spectrum (homonuclear) or the δ_x (heteronuclear).

   ¬ These spectra are particularly suitable for differentiating CH_x's (heteronuclear case) or for determining the J_{H-H}'s in a crowded region of the spectrum.

b)   J-Correlated (COSY) Spectra

   - Both homonuclear and heteronuclear cases are recognized.

   ¬ The 1-D spectrum (conventional) is along the diagonal.

   - Off-diagonal peaks show the correlation of J-couplings.

   ¬ Spectra are particularly useful for determining long range J's.

c)   δ-Correlated Spectra

   ¬ In homonuclear cases (HOMCOR) the two dimensions show chemical shifts and J's.

   ¬ In heteronuclear cases (HETCOR) the first dimension shows proton δ's and J's, while the second dimension is the chemical shift of the heteroatom, $\delta(x)$.

   - The conventional spectrum is along the diagonal.

   ¬ With HETCOR, if the heteroatom (e.g. carbon) assignments are known, one may obtain readily the proton assignments.

d)   NOESY or Nuclear Overhauser Spectra

- One dimensional spectrum is along its diagonal.

- These are particularly useful for determining distances, with off-diagonal peaks correlating nuclei close to each other.

III.G RELEVANT EXAMPLES

The preceeding sections in this chapter have dealt with the theoretical aspects of NMR pulse techniques. These methods will now be illustrated by considering the spectra of a number of fuel related samples. We will consider two types of samples. First, we will examine the spectra of n-octylbenzene, a compound that contains many of the carbon and hydrogen functionalities found in heavy fuels. Then we will consider the spectra of actual fuel samples.

III.G.1 Spectra of n-Octylbenzene

n-Octylbenzene was chosen as a model system because it contains many of the carbon and hydrogen functionalities found in heavy fuels (Petrakis, 1985). The conventional ^{1}H and ^{13}C NMR spectra of this molecule are shown in Figs. III-18 and III-19, and each of the peaks in the spectra is labeled. The ^{1}H spectrum was collected at 21°C on a Varian XL200 spectrometer operating at 200 MHz. Only four pulses were required (35° flip angle, 2 µs, 1 sec delay).

Figure III-19 is the ^{13}C spectrum. The ^{13}C spectrum was collected at 50.3 MHz with continuous broad band decoupling. The spectrum was acquired using 128 pulses. The quaternary aromatic carbon (a) is at 143 ppm, the o- and m-carbon (b,b') are at about 128 ppm and the p-carbon (c) is at 125 ppm. The signals at around 76 ppm are due to the solvent CDCℓ_3. The solvent appears as a triplet due to the ^{2}H-^{13}C coupling. The aliphatic region shows several distinct peaks: methyl carbons (k) appear at 14 ppm, (j) carbons appear at 23 ppm, (f), (g) and (h) carbons appear at 29 and 30 ppm and (i) carbons appear at 31.5 ppm and (d) carbons appear at 36 ppm.

In addition to the conventional spectra, each of the pulse sequences described in previous sections of this chapter were applied to this model compound. The PCSE spectrum is shown in Fig. III-20. A coupling constant of 140 Hz was assumed; 128 pulse repetitions were collected. The quaternary carbon (CH$_x$, x=0) at 143 ppm and all the methylene carbons (CH$_x$, x=2) between 36 and 23 ppm have signals in the positive direction, while the aromatic CH (x=1) at 125 and 128 ppm, and the methyl group (x=3) at 14 ppm have signals in the negative direction.

INEPT spectra are shown in Fig. III-21. Once again a coupling constant of 140 Hz was assumed; delay times corresponding to 1/3J, 3/4J, and 1/2J were

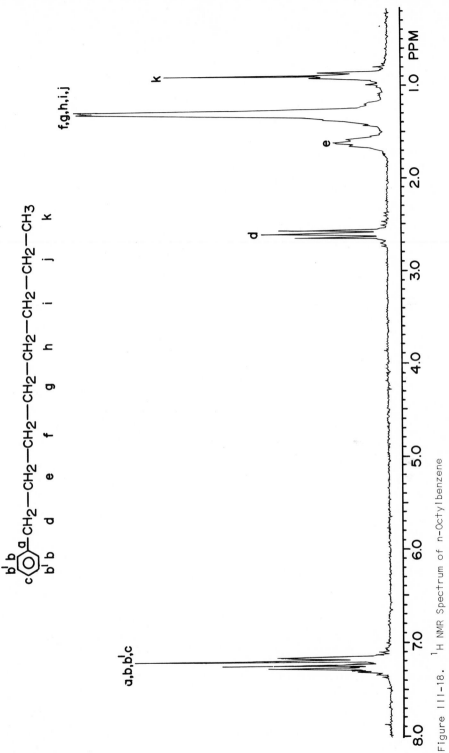

59

Figure III-18. ^{1}H NMR Spectrum of n-Octylbenzene

60

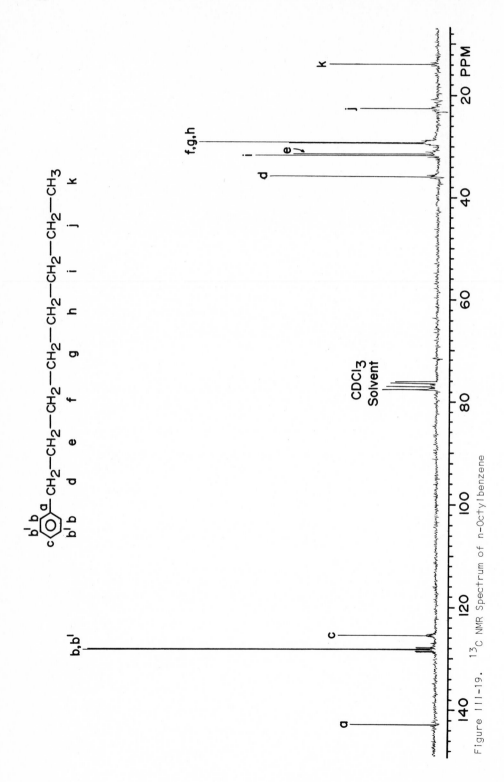

Figure III-19. ^{13}C NMR Spectrum of n-Octylbenzene

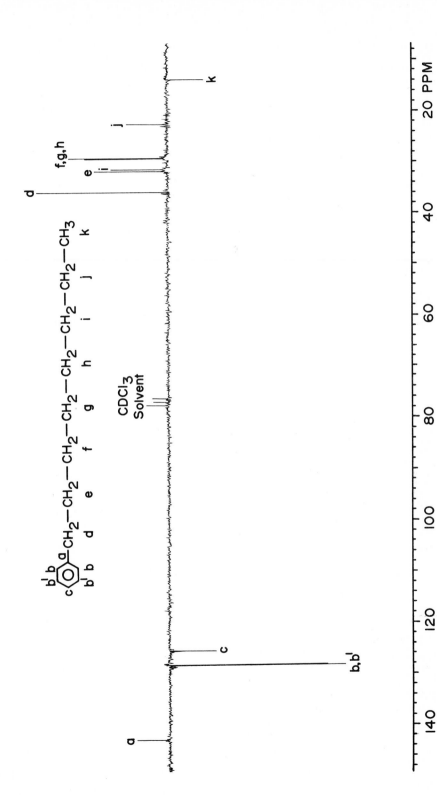

Figure III-20. PCSE Spectrum of n-Octylbenzene

61

62

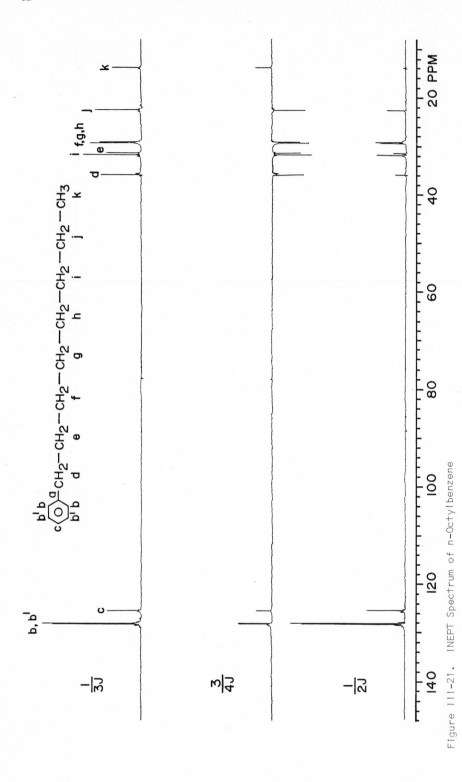

Figure III-21.   INEPT Spectrum of n-Octylbenzene

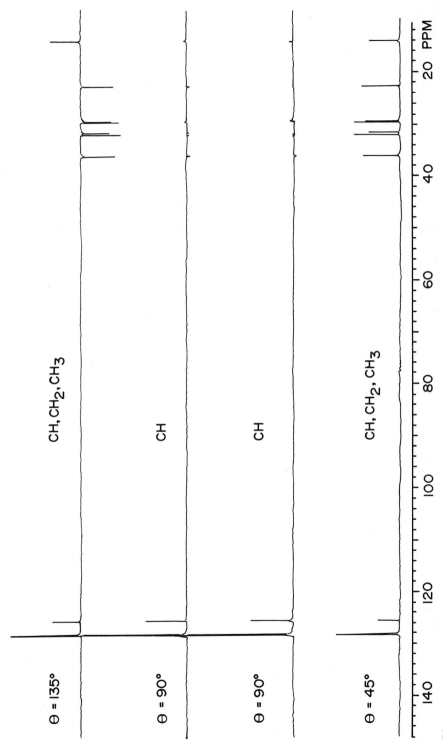

Figure III-22. DEPT Spectra for n-Octylbenzene

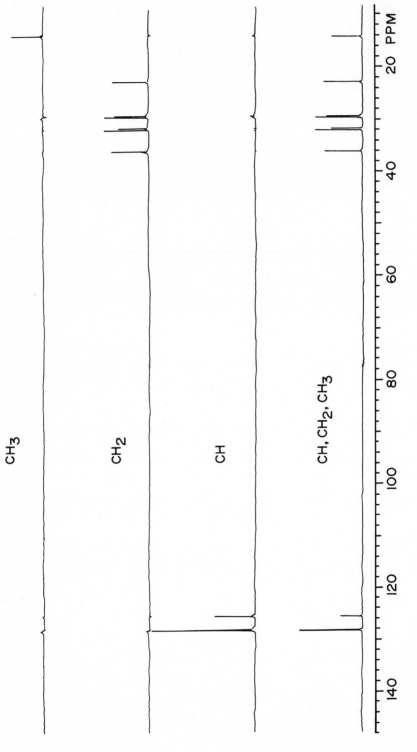

CH₃

CH₂

CH

CH, CH₂, CH₃

Figure III-23. Edited DEPT Spectra for n-Octylbenzene

used. The 1/3J spectrum is designed to create positive peaks for CH_3, CH_2, and CH carbon; quaternary carbon should not appear. The 3/4J spectrum should yield positive peaks for CH_3 and CH carbon, negative peaks for CH_2 and no quaternary carbon signal. The 1/2J spectrum should yield a signal only for CH carbon. Intercomparison of the various INEPT spectra and comparison with the conventional carbon spectrum of Fig. III-19 indicates that the pulse sequence performed largely as expected, although some residual peaks and differences in intensities can be observed (e.g. note the small CH_3 peak in the 1/2J spectrum).

DEPT results are shown in Fig. III-22. Three values for the θ pulse were used: 45°, 90° and 135°. Table III-1 indicates the expected intensities for each carbon type for these pulses. Again, some small residual peaks are observed but overall, the results are quite good. Figure III-23 shows the composite spectra, the addition/subtraction carried out according to the rules shown in Table III-2. The bottom frame is at θ = 45°, and therefore all carbons are shown. The middle two frames show only the CH and CH_2 groups. All other signals are not recorded. The top frame is, according to Table III-3, the signal of CH_2 only.

The 2-D NMR results are shown in Figs. III-24 through III-29. Figure III-24 is the HOM 2D-J spectrum, and it shows the proton chemical shifts and proton-proton coupling constants. For example, the top pattern is the methyl proton split into a triplet (J_{HH} ~ 7 Hz) by the methylene protons (j). The triplet at about 2.6 ppm is due to the protons of methylene (d) split into a triplet by the two protons of methylene. The second order pattern due to the aromatic protons is the bottom pattern at ~7.1-7.3 ppm.

The pattern in Figure III-25 shows the chemical shift of the protons which are spin-spin coupled. The existence of the symmetric, off-diagonal elements clearly indicate the existence of long-range couplings between the aromatic protons and the protons of the methylene (d), in addition to the splittings among the various aliphatic groups.

HET 2D-J (Figure III-26) shows the correlation between a given carbon and the $J_{H-^{13}C}$. At ~143 ppm on the ^{13}C chemical shift scale, there is a single peak with no splitting due to $J_{H-^{13}C}$. Clearly this is due to the quaternary carbon. The (c)-carbon (125 ppm) is split into a doublet by the lone proton to which it is directly attached, with the $J_{H-^{13}C}$ = 160 Hz, as it can be read off the spectrum directly. The pattern at 128 is due to carbons b and b', again with a $J_{H-^{13}C}$ = 160 Hz. The carbon-hydrogen splittings in the various aliphatic carbons are also obvious, e.g. the methyl carbon (14 ppm) being split into a quartet.

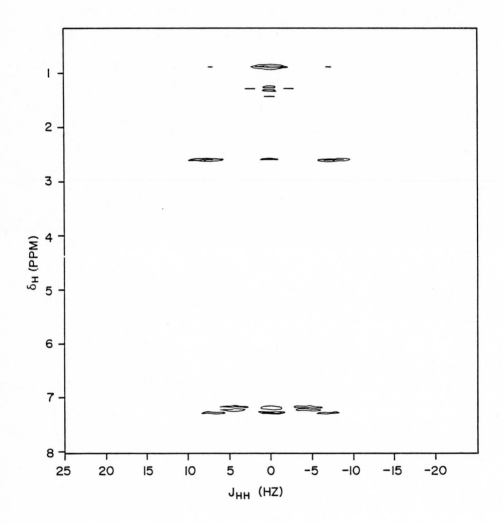

Figure III-24.   HOM2D-J ^{1}H Spectrum for n-Octylbenzene

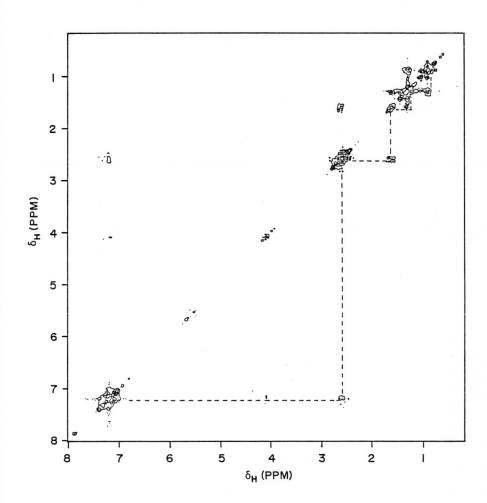

Figure III-25. HOMCOR ^{1}H Spectrum for n-Octylbenzene

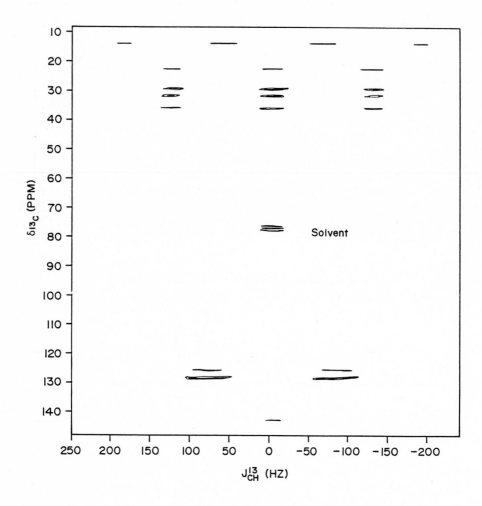

Figure III-26. HET 2D-J ^{13}C Spectrum of n-Octylbenzene

69

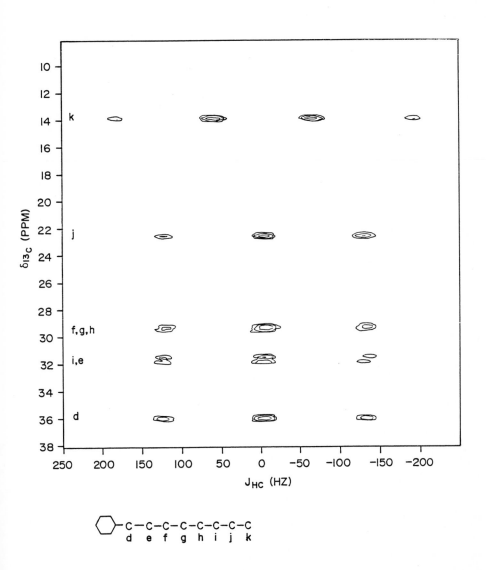

Figure III-27.   Aliphatic Region of HET 2D-J Spectrum of n-Octylbenzene

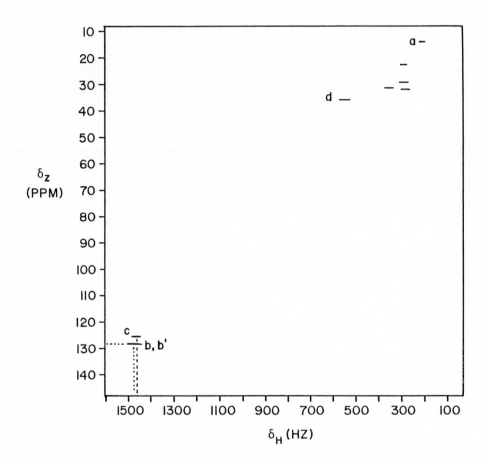

Figure III-28. HETCOR Spectrum of n-Octylbenzene

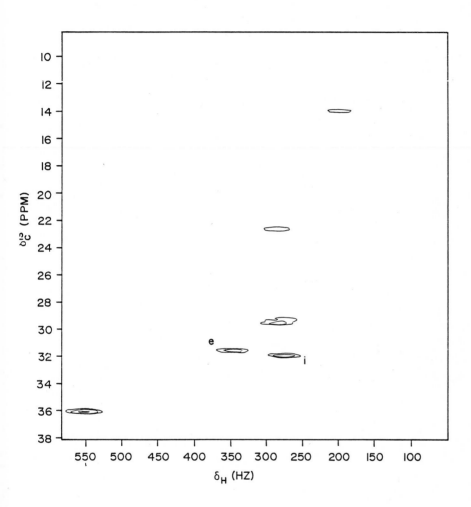

Figure III-29. Aliphatic Region of HETCOR Spectrum of n-Octylbenzene

Figure III-27 is the expanded aliphatic region from the HET 2D-J. This shows quite well the ^{13}C spectra and the $J_{H-^{13}C}$ = 125 Hz in the aliphatic region.

This HETCOR 2D spectrum (Fig. III-28) correlates the proton and carbon chemical shifts that are directly bonded to each other. The aromatic region shows only two signals c and b,b'. The quaternary is missing, since it has no directly bonded protons to it.

Figure III-29 is the aliphatic region expansion of the previous HETCOR 2D spectrum (Fig. III-28). This kind of spectrum can aid significantly in the assignment of not well-resolved carbon peaks. For example, it is not readily apparent (Fig. III-19) which of the two peaks at 31-32 ppm is i and e. Figure III-29 HETCOR resolves this question readily, for even though the two peaks are close to each other in the carbon spectra, the chemical shift difference is much more straightforward in the proton spectra.

Considering the spectra of a simple model compound such as n-octylbenzene, in which the resonance frequencies and intensities are well known, gives us some confidence in applying these techniques to less well defined systems. We will now consider the spectra of representative fuel samples.

III.G.2 Spectra of Fossil Fuels

A large number of researchers have collected conventional ^1H and ^{13}C spectra of liquid fuels. Some typical spectra have been shown in previous sections. A smaller number of investigators have applied spin echo and polarization transfer techniques to fuels. Snape (1982) has reported PCSE spectra of anthracene oils and coal derived liquids. These are shown in Fig. III-30. Dallig and coworkers (1984) have reported INEPT and J-resolved two-dimensional NMR spectra of coal derived products (Figures III-31 and III-32). Other fuel spectra using specialized pulse techniques have been reported by Schoolery, Derreppe, Gerhards and others. Some of these are shown in Figs. III-33 through III-35. The dominant feature of these spectra is the tremendous number of resonances. Without pulse techniques, deriving detailed structural information from these spectra would be difficult to do with any confidence.

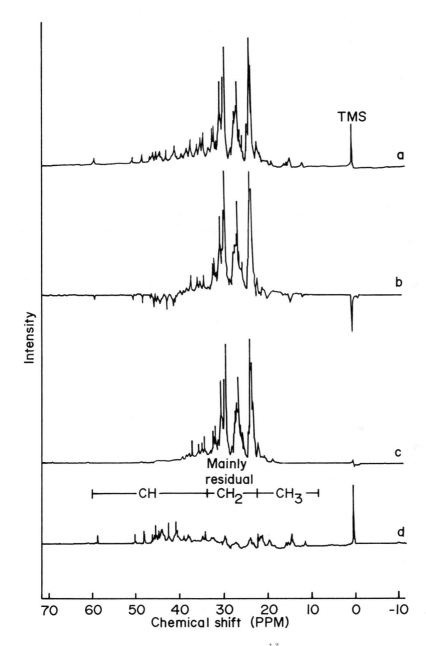

Figure III-30. Aliphatic Carbon Bands from ^{13}C NMR Spectra of Hydrogenated Anthracene Oil. (a) Normal Spectrum; (b) Positive Peaks are (C+CH$_2$), Negative Peaks are (CH+CH$_3$); (c) C+CH$_2$ Sub-spectrum; (d) CH+CH$_3$ Sub-spectrum.

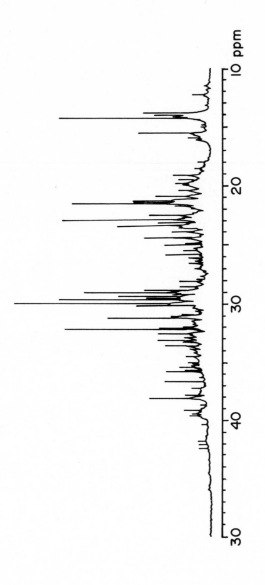

Figure III-31. Aliphatic Region of a ^{13}C NMR Spectrum of a Solvent Refined Coal-II Distillate (Dalling)

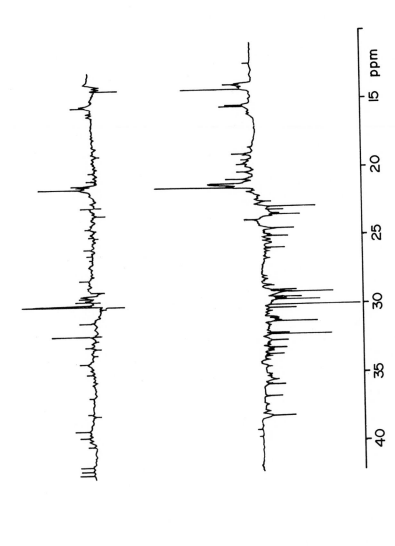

Figure III-32.   INEPT Spectra of the SRC-II Distillate of Figure III-31.   In the lower spectrum CH and CH₃ resonances are positive, CH₂ resonances are negative. In the upper spectrum, only CH resonances appear (Dalling)

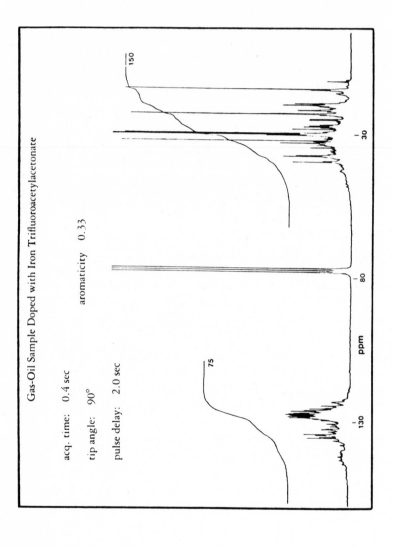

Figure III-33. ^{13}C NMR Spectrum of a Gas-Oil Petroleum Fraction (Shoolery)

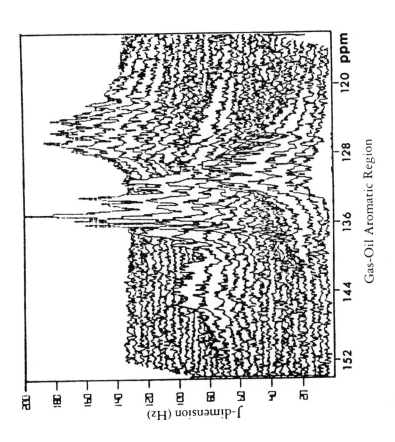

J-dimension (Hz)

Gas-Oil Aromatic Region

ppm

Figure III-34. Heteronuclear 2D J Spectrum of the Aromatic Region of a Gas-Oil Petroleum Fraction (Shoolery)

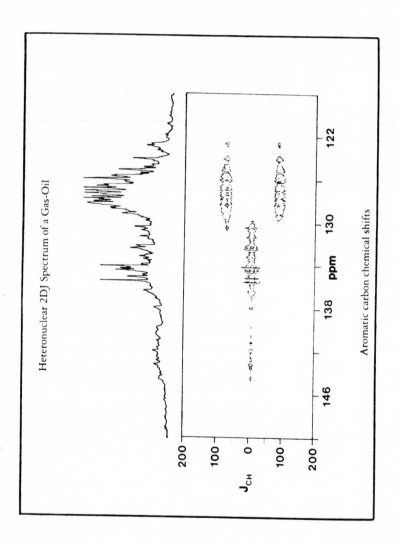

Figure III-35. Contour Plot of the Heteronuclear 2DJ Spectrum of Figure III-34 (Shoolery)

REFERENCES

Bax, A., "Two Dimensional Nuclear Magnetic Resonance in Liquids", Reidel, Boston (1982).

Bax, A., in "Magnetic Resonance: Introduction, Advanced Topics and Applications to Fossil Energy", L. Petrakis and J. Fraissard, eds., Reidel (1984) pp. 137-146

Bremser, W., Franke, E., Ernst, L., Gerhards, R. and Hardt, H., "Carbon-13 NMR Spetral Data", Verlag Chemie, Weinheim (1981).

Carrington, A. and McLachlan, A. D., "Introduction to Magnetic Resonance", Harper and Row, New York (1967).

Dallig, D. K., Haider, G., Pugmire, R., Shabtai, J. and Hidl, W., Fuel 63, 525 (1984).

Derreppe, J. M., in "Characterization of Heavy Crude Oils and Petroleum Residues", Technip, Paris (1984) pp. 298-303.

Doddrell, D. M., Pegg, D. T. and Bendall, M. R., J. Mag. Reson. 48, 323 (1982).

Gerhards, R., in "Magnetic Resonance; Introduction, Advanced Topics and Applications to Fossil Energy", L. Petrakis and J. Fraissard, eds., Reidel (1984) pp. 377-408.

Hahn, E., Phys. Today, 6, 4 (1953).

Jeener, J., Ampere International Summer School, Basko Polje, Yugoslavia (1971).

Petrakis, L. and Young, D. C., Gulf R&D Co., Pittsburgh, PA (1985).

Shoolery, Journal of Natural Products, 47(2), 242 (1984).

Snape, C. E., Fuel, 61, 775 (1982).

Soerensen, O. W. and Ernst, R. R., J. Mag. Reson., 51, 477 (1983).

PART TWO:

CHARACTERIZING LIQUID FUELS USING NMR

Chapter IV

AN OVERVIEW OF APPROACHES

IV.A INTRODUCTION

The structural characterization of complex multicomponent fractions of coal liquids, shale oils, and petroleum fractions is an important problem that has received considerable attention. Numerous separation and spectroscopic techniques have been used, either singly or in combination, in order to obtain detailed structural profiles of fossil fuels. High-resolution proton and ^{13}C NMR spectra have been particularly useful in determining the structures of the carbon skeletons of fuels. However, neither NMR spectra nor other spectroscopic and analytical data provide enough information to fully delineate the individual components present in the mixtures. Thus, recourse must be made to the approximate characterization methods of structural analysis.

One of the cornerstones of approximate characterization methods is the information provided by NMR. There are many ways of interpreting and presenting these NMR data. Of the many approaches possible, three general methods have been employed most successfully to obtain structural information from NMR spectra. These approaches are characterization by parameters, average molecule construction, and functional group analysis.

In the first of these approaches, complex hydrocarbon mixtures, such as coal liquids and petroleum distillates, are characterized by calculating a set of average structural parameters (for example, see Clutter, 1972). The parameters describe structural features, such as the fraction of carbon that is aromatic, the number and length of alkyl substituents in a ficticious average molecule, the percentage of aromatic carbons that are substituted, and the number of aromatic rings per molecule. Before ^{13}C NMR spectrometry became commonly available, the parameters were derived from ^{1}H NMR and elemental analysis data by using a number of assumptions. The nature of the assumptions depended on the specific parameters being calculated, but typical methods [Brown and Ladner, 1960] assumed a value for the average ratio of hydrogen to carbon in the aliphatic structures and that α alkyl groups have the same H/C ratio as all other alkyl groups. By utilizing the data from ^{13}C NMR spectra, many assumptions are no longer necessary and more detailed structural parameters can be determined. Given sufficient data, structural parameters can provide a useful characterization of a hydrocarbon mixture. The methods for calculating these parameters will be described in detail in a later chapter; however, the parameters approach presents two difficulties. The first difficulty is assessing the validity of the assumptions made in calculating the parameters. In most cases it is difficult, if not impossible, to check the assumptions. The second difficulty with the parameters approach is that parameters represent average values and may not provide information about the actual components of the mixture. For example, a value of 3 for the average number of carbons in an alkyl chain could indicate either a uniform distribution of chain length about 3 or a bimodal distribution with high concentrations of very short and much longer aliphatic chains.

A second approach to structural analysis utilizes NMR, elemental composition, and average molecular weight data to construct average or representative molecular structures. The structural formula of an average molecule can be determined in a straightforward manner from elemental analysis, NMR and average molecular weight data. After the structural formula is rounded to the nearest whole integers, the algorithm of Oka (1977) can be used to find all possible structures that are consistent with the analytical data. The molecular structures generated by this method are useful in visualizing the types of structures that may be present in the fuels, but they must be viewed with caution. The structures represent a statistical averaging of the properties of the molecules in the mixture and may or may not actually exist in the liquids. In addition, small variations in the average molecular formula can produce very different structures, so comparisons between similarly treated samples are difficult to make.

A third approach to the structural characterization of mixtures, termed functional group analysis, has been presented by Petrakis, Allen and coworkers (Petrakis, 1983a; Allen, 1984a,1985a). The premise of this approach is that while the number of individual molecules in a mixture may be large, all these molecules are composed of a relatively small number of functional groups. This premise is supported by a large body of experimental evidence, including results from mass spectrometry (Swansiger, 1974), IR (Solomon, 1979), NMR, and chromatography (Ruberto, 1976). Since functional groups rather than individual molecules can largely determine chemical behavior in fuel mixtures, a useful characterization of the mixture consists of listing the concentrations of the constituent functional groups. This method provides a characterization that is easy to visualize, and allows quantitative comparisons between similar samples. If sufficiently precise functional group concentrations can be determined, this method can be used to estimate thermophysical properties of the fuels and to elucidate reaction pathways and kinetics.

In the following chapters we will describe each of the methods of structural characterization in detail. Sample calculations will be performed and the results of the various methods will be compared. In addition, we will briefly examine the uses of such structural characterization in modeling the thermophysical properties of fuels. Although all of the examples will deal with fuels, it is important to note that these methods are applicable to any complex mixture. Before proceeding on to this discussion, however, we will briefly sample the literature on NMR as applied to liquid fuels.

IV.B REPRESENTATIVE APPLICATIONS

High resolution NMR has been used extensively and with great success in the study of liquid fuels. Many of the early studies involved ^{1}H NMR, but even before the advent of commercially available ^{13}C NMR, it was recognized that the study of the liquid fuels was best through the carbon skeleton, despite the difficulties of working with the insensitive ^{13}C isotope. With the recent wide availability of FT NMR, ^{13}C NMR, along with the experimentally easier ^{1}H NMR, has come to be a most powerful structural tool. ^{17}O, ^{15}N, ^{29}Si, ^{19}F and other nuclei have also found use as promising probes of the structure of liquid fuels and of their changes upon processing. Not surprisingly, the relevant literature has grown immensely in the last several years. The survey given here is not exhaustive, but rather it is meant to be a representative sampling and an indication of the power of the NMR technique.

Conference Proceedings

- A 1985 ACS meeting had a symposium devoted to magnetic resonance techniques applied to heavy ends (Retcofsky and Rose, 1985)

- An international symposium on the Characterization of Heavy Crude Oils and Petroleum Residues (Lyon, 1984) summarizes much of what is known from NMR about the chemistry and structure of these materials.

- Petrakis and Fraissard (1984) have edited the proceedings of an advanced study institute devoted to magnetic resonance and its many applications to fossil energy problems. Quantitative NMR, NMR pulse techniques, very high field NMR and ^{17}O NMR are but a few of the many topics addressed.

- Cooper and Petrakis (1981) edited the proceedings of an American Institute of Physics Conference on coal utilization. Many analytical techniques used to study fossil fuels are addressed.

Reviews

- C. Karr (1978) has compiled a four volume treatise devoted to essentially all aspects of fossil fuel analyses. This work reviews most of the literature prior to 1978.

- The newly revised supplement to Lowry's treatise on coal chemistry (Elliot, 1981) contains a review of NMR techniques.

- The utilization of NMR in petroleum, coal and oil shale analysis has been reviewed by L. Petrakis and E. Edelheit (1979). The review addresses applications in the entire range of fuel research, from exploration to environmental aspects.

- A multivolume work on the characterization of coal liquid products has been edited by H. D. Schultz (1983).

- An issue of the journal "Liquid Fuel Technology" has been devoted to structural characterization techniques (Speight, 1984).

Applications

We have separated our sampling of research papers on fuel charcterization using NMR into three sections. One section is devoted to coal derived materials, another to shale oils and the third to heavy oils. The recent literature on coal derived materials is the most extensive, but heavy oils are also receiving much attention. It should be noted that most of the techniques described in these papers are not restricted to a single type of fuel. Rather the techniques are generally applicable to any liquid fuel.

Shale oils:

- Analyses of shale oil components by high resolution NMR have been performed by Netzel [1980,1981].

- Miknis [1982], Parks [1985] and Lambert [1985] have addressed NMR applications in oil shale processing.

- 2-D NMR techniques have been applied to oil shales and tar sands (Sobol et al., 1985).

Coal derived materials:

- Whitehurst (1982) has used field ionization mass spectroscopy (FIMS) in combination with NMR and other analytical techniques to characterize coal liquids.

- NMR has been used to study the structure of coal extracts by a number of investigators, including the early seminal work by Bartle, Martin and Williams (1975) and Ladner and Stacey (1961).

- NMR has been used as a structural probe for coal liquefaction and pyrolysis products by scores of researchers. Some representative publications include those by Snape (1982), Clarke (1982), Yokoyama (1979), and Gavalas and Oka (1978).

- Pulse techniques have been used by Snape (1985), Dalling (1984), and others to enhance the characterization of coal liquids.

- High molecular weight coal derivatives have been characterized using NMR (Bartle and Zander, 1983; Bartle and Jones, 1983).

- Young (1982,1983) has used NMR to characterize upgraded products from the SRC and H-Coal liquefaction processes.

- The structure of coal derived asphaltenes as a function of molecular mass have been examined by Richards (1983).

Heavy oils:

- NMR has been used to "fingerprint" crudes (Bouquet and Bailleul, 1982).

- Gillet (1981) has reviewed the criteria for obtaining reliable structural parameters for oil products using NMR.

- Cookson and Smith (1983a,1983b) have used off-resonance decoupling techniques to discriminate between CH, CH_2, and CH_3 carbons in oils.

- Holak (1984) has used both chemical shifts and spin-lattice relaxation times to characterize oil products.

- The nitrogen compounds present in crude oils have been examined using NMR (Khan, 1982).

This is but a sampling of the extensive literature available. Fuel and preprints of the Fuel Division and Petroleum Division of the American Chemical Society continue to be important vehicles for publication of specialized results.

REFERENCES

Allen, D. T., Petrakis, L., Grandy, D. W., Gavalas, G. R., and Gates, B. C., Fuel, 63, 803 (1984a).

Allen, D. T., Grandy, D. W., Jeong, K. M., and Petrakis, L., Ind. Eng. Chem. Process Des. Dev., 24, 737 (1985a)

Bartle, K. D., Martin, T. G., and Williams, D. F., Fuel, 54, 226 (1975).

Bartle, K. D., and Zander, M., Erdoel Kohle, Erdgas, Petrochem., 36(1), 15 (1983).

Bartle, K. D., and Jones, D. W., "TrAC, Trends Anal. Chem., 2(6), 140 (1983).

Bouquet, M., and Bailleul, A., Proc. Inst. Pet., London, 2, 394 (1982).

Brown, J. K. and Ladner, W. R., Fuel, 39, 87 (1960).

Clarke, J. W., Rantell, T. D., and Snape, C. E., Fuel, 61, 707 (1982).

Clutter, D. R., Petrakis, L., Stenger, R. L., and Jensen, R. K., Anal. Chem., 44, 1395 (1972).

Cookson, D. F., and Smith, B. E., Fuel, 62, 34 (1983a).

Cookson, D. J., and Smith, B. E., Fuel, 62, 39 (1983b).

Cooper, B. R., and Petrakis, L., Editors, Proceedings AIP Conference, No. 70, American Institute of Physics, New York (1981).

Dalling, D. K., Haider, G., Pugmire, R., Shabtai, J. and Hull, W., Fuel, 63, 525 (1984).

Elliot, M., "Chemistry of Coal Utilization", Suppl. Vol. (Rev.), Wiley, New York (1981).

Gavalas, G. R., and Oka, M., Fuel, 57, 285 (1978).

Gillet, S., Rubini, P., Delpuech, J. J., Escalier, J. C., and Valentin, P., Fuel, 60, 221 (1981).

Holak, T. A., Aksnes, D. W., and Stocker, M., Anal. Chem., 56, 725 (1984).

Karr, C., Jr. (editor), "Analytical Methods for Coal and Coal Products", (four volumes), Academic Press, 1978.

Khan, I. A., Al-Asadi, Z. A. K., Muttawalli, F. S., and Tameesh, A. H., J. Pet. Res., 1(1), 58 (1982).

Ladner, W. R., and Stacey, A. E., Fuel, 40, 119 (1961).

Lambert, D. E. and Wilson, M. A., Prepr. Am. Chem. Soc. Div. Fuel Chem., 30(3), 256 (1985).

Lyon, International Symposium on the Characterization of Heavy Crude Oils and Petroleum Residues, Technip, Paris (1984).

Miknis, F. P., Proc. Intersoc. Energy Convers. Eng. Conf., 17(2), 935, 1982.

Mitchell, P. C. H., Read, A. R., Calclough, T., and Spedding, H., Proc. Int. Conf. Chem. Uses Molybdenum, Ann Arbor, Mich., 1982.

Netzel, D. A., Institute of Gas Technology Meeting, 271 (1980).

Netzel, D. A., McKay, D. R., Heppner, R. A., Guffey, F. D., Cooke, S. D., Varie, D. L., and Linn, D. E., Fuel, 60, 307 (1981).

Oka, M., Chang, H., and Gavalas, G. R., Fuel, 56, 3 (1977).

Parks, T., Lynch, L. J. and Webster, D. S., Prepr. Am. Chem. Soc. Div. Fuel Chem., 30(3), 247 (1985).

Petrakis, L., Allen, D. T., Gavalas, G. R., and Gates, B. C., Anal. Chem., 54, 1557 (1983a).

Petrakis, L., and Fraissard, J. (Editors), "Magnetic Resonance: Introduction, Advanced Topics and Applications to Fossil Energy", Reidel (Dordrect/Boston), 1984, 807 pp.

Petrakis, L., and Weiss, F. T. (Editors), "Petroleum in the Marine Environment", ACS Advances in Chemistry Series No. 185, 1981.

Petrakis, L., and Edelheit, E., Appl. Spectrosc. Rev., 15(2), 195 (1979).

Retcofsky, H., and Rose, K. D., Prepr. Am. Chem. Soc. Div. Petr. Chem., 30(2), 232 (1985).

Richards, D. G., Snape, C. E., Bartle, K. D., Gibson, C., Mulligan, M. J., and Taylor, N., Fuel, 62, 724 (1983).

Ruberto, R. G., Jewell, D. M., Jensen, R. K., and Cronauer, D. C., Adv. Chem. Ser. No. 151, Chapter 3, 1976.

Schultz, H. D. (editor), "NMR Spectroscopic Characterization and Production Processes", 2 volumes, Wiley, NY, 1983.

Snape, C. E., Fuel, 61, 775 (1982).

Snape, C. E. and Marsh, M. K., Prepr. Am. Chem. Soc. Div. Petr. Chem., 30(2), 247 (1985).

Sobol, W. T., Schreiner, L. J., Miljkovic, L., Marcondes-Helene, M. E., Reeves, L. W., and Pintar, M. M., Fuel, 64, 583 (1985).

Solomon, P. R., Prepr. Am. Chem. Soc. Div. Fuel Chem., 24(2), 184 (1979).

Speight, J., (ed.) Liquid Fuels Technology, 2(3), 211 (1984).

Swansiger, J. T., Dickson, F. E., and Best, H. T., Anal. Chem., 46, 730 (1974).

Whitehurst, D. D., Butrill, S. E., Derbyshire, F. J., Farcasiu, M., Odoerfer, G. A., and Rudnick, L. R., Fuel, 61, 994 (1982).

Yokoyama, S., Bodily, D. M., and Wiser, W. H., Fuel, 58, 162 (1979).

Young, L. J. S., Li, N. C., and Hardy, D., Prepr. Am. Chem. Soc., Div. Fuel Chem., 27(3-4), 139 (1982).

Young, L. J. S., Li, N. C., and Hardy, D., Fuel, <u>62</u>, 718 (1983).

Chapter V

AVERAGE MOLECULAR PARAMETERS

V.A   METHODS OF CALCULATING AVERAGE MOLECULAR PARAMETERS

The method of characterizing fuel fractions by calculating values of structural parameters was the first method of structural characterization to appear
in the literature. The procedure arose out of the need to quantitatively
describe the structure of petroleum fractions (Williams, 1958; Knight, 1967;
Brown and Ladner, 1960; Clutter et al., 1972).

Table V-1 lists and describes the average parameters which are generally
determined for petroleum fractions. Various approaches can be taken to determine these parameters. Some methods require only NMR data while others
require information such as elemental analysis and an average molecular weight
of the sample, in addition to NMR. It is important to realize that these
average parameters are the weighted average for certain properties of the sample and may or may not represent actual structures present in the fuels.   In
most instances the values determined are non-integral.

Several calculation methods have been used to determine average parameters
of petroleum fractions. The following is a brief review of existing methods.

V.A.1 ^{1}H NMR Methods

The first of the structural characterization methods based on ^{1}H NMR was
originally described by Brown and Ladner (1960) (Method 1). This method uses
the proton magnetic resonance spectrum and C and H elemental analysis data to
determine the fraction of carbon that is aromatic, i.e., the aromaticity of a
sample. The normalized integrated intensities of the types of hydrogen are
determined and used in equation 1 to calculate the fraction of carbon that is
aromatic.

$$f_a(1) = \left[(C/H)-(H_\alpha^*/x) - \frac{(H_\gamma^*+H_\beta^*)}{y} \right]/(C/H) \tag{1}$$

where $f_a(1)$ is the carbon aromaticity derived using method 1, C/H is the
atomic ratio of carbon/hydrogen from an elemental analysis, x and y are the
average number of hydrogens per α-alkyl and the remaining alkyl, respectively.
H_α^* and $(H_\gamma^*+H_\beta^*)$ represent the normalized integrated intensities of the α-alkyl
and other alkyl hydrogens, respectively. The resonance positions of the vari-
ous hydrogen types are shown in Figure V-1. Using equation 1, the aromaticity

92

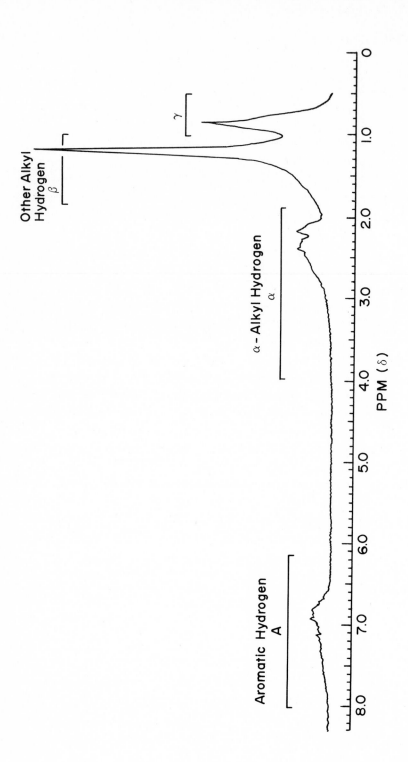

Figure V-1. ^{1}H NMR Band Assignments Used in Calculating Aromaticity

Table V-1. Average Parameters Generally Calculated for Liquid Fuels

n = average number of carbon atoms per alkyl substituent.

f = average carbon-hydrogen weight ratio of the alkyl groups.

$\%AS$ = per cent substitution of alkyl groups on non-bridge aromatic ring carbons.

$\#C_A$ = average number of aromatic ring carbon atoms per average molecule.

$\#C_1$ = average number of non-bridge aromatic ring carbon atoms per average molecule.

R_A = average number of aromatic rings per average molecule.

R_S = average number of alkyl substituents per average molecule.

R_N = average number of naphthenic rings per average molecules.

f_a = molar ratio of aromatic carbon to total carbon in sample.

BI = branchiness index.

MW = average molecular weight.

$\%C_S$ = percentage saturate carbon.

$\%C_N$ = percentage naphthenic carbon.

r = number of naphthalene rings per substituent.

$\%C_1$ = fraction non-bridge aromatic ring carbons.

of a fuel having a molar carbon to hydrogen ratio of 1.0 and 20% alpha hydrogen, 70% beta hydrogen and 10% gamma hydrogen would be estimated to be

$$f_a(1) = \left[1.0-(0.2/x) - \frac{(0.1+0.7)}{y}\right]/1.0$$

The problem which arises in using this method is the estimation of the parameters x and y. Brown and Ladner (1960) performed studies to estimate these parameters using coal-like materials and concluded that $1.5 < x,y < 2.5$. The values usually assumed for x and y are 2.0.

Petrakis and coworkers have performed experiments to ascertain the sensitivity of $f_a(1)$ upon variation of x and y. In order to simplify the problem x was set equal to y. Figure V-2 shows the result of changing x and y in steps of 0.1 from 1.8 to 2.2. The four lines correspond to the four steps; 1.8 to 1.9, 1.9 to 2.0, etc. This plot shows that the change in f_a is a function of both f_a and x. The sensitivity of f_a is approximately 0.03 unit in the region of most aromatic fractions of conventional heavy petroleum distillates. This indicates an absolute deviation of approximately 0.03 for a change of 0.1 in the values of x or y.

Carbon aromaticity is the only average parameter calculated by this method. As a result, the information concerning a sample is limited when Method 1 is used to determine average parameters. Despite the fact that f_a can be measured directly using ^{13}C NMR and despite the uncertainty involved in applying this method, equation 1 is still frequently used.

A second ^1H NMR method, as originally presented by Williams (1958), involves a detailed treatment of an aromatic fraction of an oil sample. The information which is required to obtain average parameters by this method are the integrated intensities of the proton spectrum, carbon and hydrogen elemental analysis, average molecular weight, and a "branchiness index" (see below). There are two assumptions that must be made to perform this analysis. First, the carbon to hydrogen ratio of the alkyl groups needs to be accurately estimated. This involves the determination of a "branchiness index". Second, the carbon to hydrogen ratio of the α-alkyl groups is assumed equal to that of the other alkyl groups.

The equations used to calculate the various parameters estimated by this method are listed in Table V-2. Equation 2 involves the second assumption stated above and equation 3 involves the first assumption.

The "branchiness index", BI, that is used in equation 4 and indirectly in all the other equations except equation 2, is defined as the peak height ratio of the gamma to beta protons (Williams, 1958). The argument is that as the amount of naphthenic material in the sample increases, this branchiness index

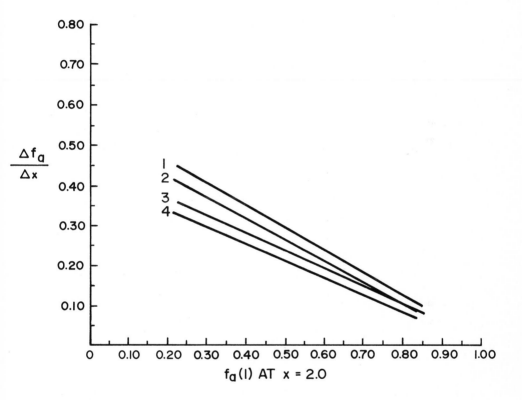

Figure V-2. Effect of the Parameters x and y on the Aromaticity Calculated
Using the Brown and Ladner Equation; Line 1 Represents the Effect
of Increasing x and y from 1.8 to 1.9; Line 2 Represents the
Effects of Increasing x and y from 1.9 to 2.0, etc.

Table V-2. Equations Used in the ^{1}H NMR Method of Williams (1958)*

$$n = (H_\alpha^* + H_\beta^* + H_\gamma^*)/H_\alpha^* \tag{2}$$

$$f = 12n/(2n+1-2r) \tag{3}$$

$$r = \frac{(0.250[BI+4.12]-1)}{2} \ (n-1) \tag{4}$$

$$C_A = C-C_S \tag{5}$$

$$C_S = f(H_\alpha^* + H_\beta^* + H_\gamma^*)H \tag{6}$$

$$f_a(2) = C_A/C \tag{7}$$

$$C_1 = (12.0\ H_A^* + fH_\alpha^*)H \tag{8}$$

$$\#C_1 = C_1 MW/1200 \tag{9}$$

$$\%AS = (fH_\alpha^* H)100/C_1 \tag{10}$$

$$\#C_A = C_A MW/1200 \tag{11}$$

$$R_A = \frac{\#C_A - \#C_1}{2} + 1 \tag{12}$$

$$R_N = (\%AS)(\#C_1)r/100 \tag{13}$$

$$R_S = (\%AS)(\#C_1)/100 \tag{14}$$

*Variables are defined in Table V-1. Additional variables include

C_A = weight percentage of aromatic carbon

C_S = weight percentage of saturate carbon

C_1 = weight percentage of C_1 carbon

H = weight percentage of hydrogen

will increase proportionally. This argument is questionable from an experimental point of view, in that several of the terminal methyl groups in naphthenic material have resonance positions in the region of methylene groups. This question could be resolved by detailed analysis of narrow petroleum cuts by carbon-13 magnetic resonance.

The molecular weight used in equations 9 and 11 is generally determined by either vapor phase osmometry (VPO) or mass spectrometry. Figure V-3 shows the correlation between the VPO and mass spectrometry molecular weights for a series of FCC charge stocks. Theoretically, the points in the graph should fall on the dotted line, but as the molecular weight increases, the two methods disagree, probably due to the fact that the VPO molecular weight is a number average rather than a weight average molecular weight. Thus, depending upon the molecular weight range and the method of determining the molecular weight, there may be variations in the values of some of the structural parameters.

Two clarifying remarks may be needed here. First, in the definitions of the average number of carbon atoms per substituent (n) and the average number of alkyl groups per molecule (R_s), it is assumed that a system such as tetralin (1,2,3,4 tetrahydronaphthalene) has two alkyl groups of two carbons each. Second, no allowances are made in the preceding formulations for a system with more than one aromatic nucleus per molecule. Thus, if a large percentage of the molecules in the sample were biphenyls, some of the average parameters would be in error. Williams (1958) presents some empirical expressions to account for this situation, but he states that for systems with molecular weights less than 400, the error introduced by not considering the above types of molecules is minimal.

A third ^{1}H method for characterizing fuels has been presented by Clutter et al. (1972). In this method, the aromatic fraction of a petroleum sample is characterized through a detailed analysis of its proton magnetic resonance spectrum. Given a proton magnetic resonance spectrum, this method can be used to calculate all of the average parameters determined by Williams' method, the fraction of monoaromatic and diaromatic (fused) ring systems, the average molecular formula, and the average molecular weight. No information other than ^{1}H NMR is necessary to perform these calculations. In order to extract these average parameters from the proton magnetic resonance spectrum, certain assumptions concerning the sample are required. The first assumption, which limits the size of ring systems considered, is that the sample contains only mono- and diaromatic (fused) ring systems. The validity of this assumption will vary with the volatility of the sample. For most distillable petroleum fractions, separation data supports the validity of this assumption (Jewell, 1972a,1972b).

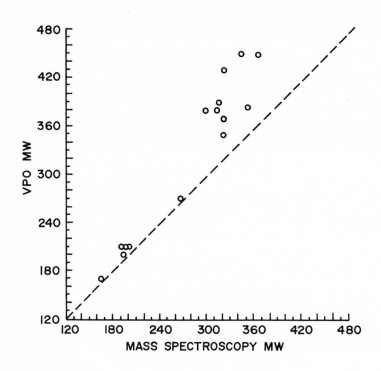

Figure V-3. Comparison of Molecular Weights Obtained Using VPO and MS

The second assumption is concerned with the spectral interpretation. It requires that the monoaromatic and diaromatic ring protons can be separated in the proton magnetic resonance spectrum. This assumption is supported by a survey of selected NMR spectra of mono- and diaromatic systems.

The third assumption is concerned with the number of substituents in the mono- and diaromatic components. There are two alternative assumptions that can be made: the number of substituents is the same in both components; the percentage of substitution is the same in both components. By observation of a number of mono- and diaromatic cuts, the first assumption appears to be more reasonable. The percentage substitution of monoaromatics has been observed to be greater than 50% while that for the diaromatics is less than 50%.

The equations used in this method are those listed in Table V-3 along with equations 2, 12 and 14 from Table V-2. The first parameter calculated is the fraction monoaromatic rings in the sample by equation 25, where H_A^m and H_A^D represent the intensity of the mono- and diaromatic protons, respectively. H_A^D is that portion of the aromatic integrated intensity below 7.05 ppm and H_A^m is that portion of the aromatic integrated intensity above 7.05 ppm. The factor (6-#α-alkyl carbons)/(8-#α-alkyl carbons) is required to yield the proper ratio since the number of non-bridge aromatic ring carbons is not the same in mono- and diaromatic ring systems, and the percentage substitution is not the same. Once $\#C_1$ is known, then the number of alkyl carbons can be calculated since the relationship between intensity units and number of carbon atoms is now defined. In the calculation of monoaromatics, an iterative process using equations 25 to 31 is used since the number of α-alkyl carbons is not known initially. The value of the variable mono normally converges in approximately two iterations. From equation 35, $\#C_1^u$ represents the number of aromatic protons observed in the spectrum, thus the total number of hydrogens in the average molecule, total #H, can be calculated from the relationship between aromatic intensity units and number of aromatic hydrogen atoms.

The amount of naphthenic carbon is determined in the following manner. Naphthene rings fused to aromatic systems yield proton NMR spectra with an absorption between 1.65 and 1.9 δ which is assigned to the β-naphthenic protons on the unsubstituted carbons. The protons on substituted β-naphthenic carbons yield a resonance slightly upfield from this region in most cases. Thus, to determine the average number of naphthenic carbons, one-half of the region from 1.65 to 1.9 δ corresponds to the number of unsubstituted β-naphthenic carbons and assuming that an equal number of substituted β-naphthenic carbons are present, the total number of naphthenic carbons is easily determined. Then the per cent naphthenic carbon ($\%C_N$) is just given by equation 41. The average number of naphthenic rings (R_N) is calculated by

Table V-3. Equations Used in the ^{1}H NMR Method of Clutter (1972)

$$\text{mono} = \frac{H_A^m}{H_A^m + H_A^D \left[\dfrac{6 - \#\alpha \text{ alkyl carbons}}{8 - \#\alpha \text{ alkyl carbons}}\right]} \tag{25}$$

$$\text{di} = 1 - \text{mono} \tag{26}$$

$$\#\text{mono} = 6 \text{ mono} \tag{27}$$

$$\#\text{di} = 10 \text{ di} \tag{28}$$

$$\#\text{bridge} = 2 \text{ di} \tag{29}$$

$$\#C_A = \#\text{mono} + \#\text{di} \tag{30}$$

$$\#C_1 = \#C_A - \#\text{bridge} \tag{31}$$

$$\text{Total } \#C = \#C_A + \text{number alkyl carbons} \tag{32}$$

$$f_a(3) = \#C_A / \text{total } \#C \tag{33}$$

$$\%AS = 100(\#\alpha \text{ alkyl carbons}) / \#C_1 \tag{34}$$

$$\#C_1^u = \left(1 - \frac{\%AS}{100}\right) \#C_1 \tag{35}$$

$$\text{av molecular formula} = C_{(\text{total } \#C)} H_{(\text{total } \#H)} \tag{36}$$

$$\text{av molecular weight} = 12.01 \text{ (total } \#C) + 1.008 \text{ (total } \#H) \tag{37}$$

$$f = \frac{12.01(\text{total } \#C - \#C_A)}{1.008(\text{total } \#H - \#C_1^u)} \tag{38}$$

$$\%C_S = \frac{100(\text{total } \#C - \#C_A)}{\text{total } \#C} \tag{39}$$

$$\%C_1 = \frac{100 \#C_1}{\text{total } \#C} \tag{40}$$

$$\%C_N = \frac{100 \ \#\text{naphthenic carbons}}{\text{total } \#C} \tag{41}$$

$$R_N = \frac{\#\text{naphthenic carbons}}{3.5} \tag{42}$$

equation 42, where 3.5 assumes an equal amount of five- and six-membered naphthenic rings. Although this is strictly an estimate of the per cent naphthenic carbon, the value should reflect major changes in composition from sample to sample. A FORTRAN program for performing these calculations is given in the Appendix.

The final ^{1}H NMR method that we will present is just a graphical extension of Brown and Ladner's equation. The aromaticity value determined in equation 1 is a function of both the integrated intensities and the C/H ratio. Figure V-4 shows the relationship between f_a and $(H_\alpha^* + H_\beta^* + H_\gamma^*)/2$ constructed from a number of analyses of petroleum fractions. The figure indicates that it is possible to graphically obtain the aromaticity of a petroleum fraction without knowledge of the C/H ratio. This method is obviously limited and assumes that the C/H ratio is the same for any given $(H_\alpha^* + H_\beta^* + H_\gamma^*)/2$ regardless of the composition of the sample.

V.A.2 ^{13}C NMR Method

Through the combination of carbon-13 and proton magnetic resonance, Knight (1967) developed a scheme to determine the average parameters of aromatic fractions of petroleum in a manner quite similar to the method of Williams. Essentially, the same average parameters are determined, but information from the carbon-13 spectrum is used whenever possible.

In order to ascertain the average molecular parameters by the carbon-13 method, the integrated intensities of the characteristic resonances in both the carbon-13 and proton magnetic resonance spectra are required, along with elemental analysis and the average molecular weight. The only assumption involved in this technique is that the carbon to hydrogen ratio of the α-alkyl and other groups is the same (second assumption in Williams' method).

From the carbon-13 magnetic resonance spectrum of a petroleum fraction, the aromaticity of the sample can be determined directly from the integrated intensities. For this method, three characteristic regions are discernible with relative areas A_1, A_2 and A_3 as is shown in Figure V-5. Other methods derive more information from the carbon spectrum, as will be discussed later. For this method, the region A_1, characterizes ring junction carbon, substituted ring carbon, and one-half of the unsubstituted ring carbon. The next region, A_2, indicates the other one-half of the unsubstituted ring carbon, and region A_3 contains the resonances of the carbon in saturated groups.

The equations used in this method are those listed in Table V-4 along with equations 2, 9, 11, 12 and 14 from Table V-2. The average number of carbon atoms per saturated substituent (n) is calculated by equation 2. Equations 18 and 19 are alternative ways to calculate C_1^u. Normally, equation 19 is

102

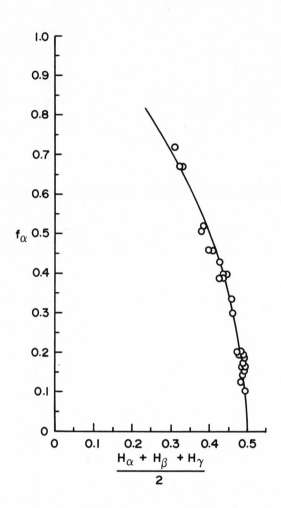

Figure V-4. Graphical Method for Estimating Carbon Aromaticity

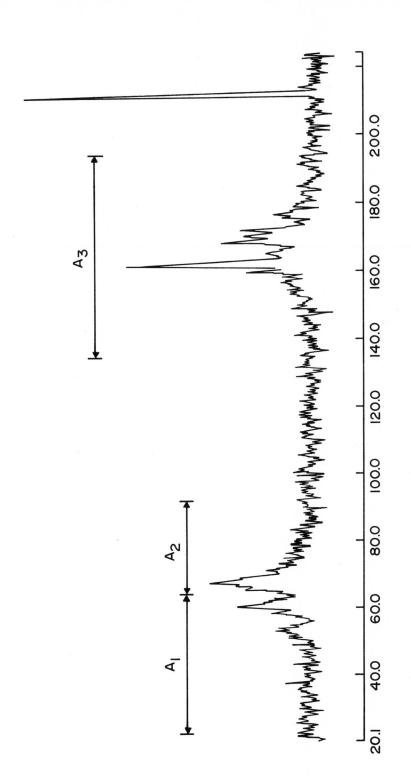

Figure V-5. ^{13}C NMR Band Assignments Used in Calculating Average Parameters

Table V-4. Equations Used in Carbon-13 Method

$$f_a(^{13}C) = (A_1+A_2)/(A_1+A_2+A_3) \qquad (15)$$

$$\#C_a = f_a(^{13}C)\text{total } \#C \qquad (16)$$

$$\#C_1^s = A_3\text{total } \#C/n \qquad (17)$$

$$\#C_1^u = 2A_2\text{total } \#C \qquad (18)$$

$$\text{total } \#C = (C/1200)\times MW \qquad (19)$$

$$\#C_1 = \#C_1^s+\#C_1^u \qquad (20)$$

$$\%AS = \#C_1^s100/\#C_1 \qquad (21)$$

$$f = A_3C/(H_\alpha^*+H_\beta^*+H_\gamma^*)H \qquad (22)$$

$$r = (n+0.5)-(6n/f) \qquad (23)$$

$$R_N = R_S r \qquad (24)$$

recommended for this calculation since there is a minimum of overlap in this resonance region. In this method, r is not estimated from the proton branchiness index as in the method of Williams but is calculated by equation 23 from n and f.

Thus, these average parameters are based upon the direct observation of the carbon skeleton via the carbon-13 magnetic resonance. In addition to being able to directly measure the aromaticity, this method represents a reliable means of estimating the number of naphthenic rings per average molecule.

V.A.3 Comparison of ^{1}H and ^{13}C Methods

The Brown and Ladner method (which we will call method 1) and the graphical extension represent one type of analysis whereas Williams' method (method 2), Clutter's method (method 3) and the ^{13}C method (method 4) represent another. Therefore, for comparison purposes, Method 1 and the graphical method are grouped together as are Method 2, Method 3, and Method 4 (carbon-13).

Since the graphical method is essentially identical to Method 1, the results provided by these two methods should be equivalent if the C/H ratio varies in a predictable manner. As a result, the graphical method yields a value of the aromaticity as soon as the proton spectrum can be recorded, whereas Method 1 requires the elemental analysis. Therefore, for a rapid indication of the aromaticity, the graphical method is the method of choice.

If average parameters other than aromaticity are desired, then Methods 2, 3, and 4 (carbon-13) must be considered. Methods 2 and 4 (carbon-13) parallel each other somewhat in determining the average parameters, whereas Method 3 is a result of a different approach. Upon comparing Methods 2 and 4, one sees that the questionable assumption of a "branchiness index" is not necessary when carbon-13 data are available. Therefore, the carbon-13 method is preferred over Method 2 if ^{13}C data are available.

Variations in structural parameters determined by method two and the ^{13}C method due to changes in carbon and hydrogen elemental analysis have been determined. For method two there is an approximate decrease of 0.06 unit in the aromaticity for a positive increase of 1% in the weight per cent hydrogen. Other average parameters which decrease with the increase in weight per cent hydrogen are $\#C_A$ and R_A, and an increase in $\#C_1$, R_S, and R_N is observed. Thus, very careful and accurate elemental analyses are required for both Method 2 and the carbon-13 method to provide accurate average parameters of the sample.

In the case of Method 3, no experimental errors other than the quality and measurability of the spectrum are involved. The errors introduced from the spectral measurements of the proton spectrum reflect a variation of approximately +0.01 in the aromaticity and an error of ±0.03 in the aromaticity from the carbon-13 spectrum.

No further absolute comparisons are possible for the different methods since there is no technique at this time which yields the exact composition of a petroleum fraction. In order to further ascertain the quality of results from each method, the actual results of each method must be compared with each other and with the spectral features of the proton and carbon-13 magnetic resonance spectra. From this type of comparison, the technique which best reflects the differences or changes in the composition of a petroleum fraction can be ascertained. We now present an experimental comparison of these methods.

All of the methods described previously have been applied to the problem of characterizing the aromatic fraction of FCC charge stocks (Clutter et al., 1972). Table V-5 shows the aromaticities of four such samples determined by the five methods. Note that the carbon-13 method involves the fewest assumptions and is the method of directly counting the aromatic and aliphatic carbons, whereas all other methods calculate the carbon aromaticity by indirect means. Table V-5 indicates that Method 3, carbon-13, and the graphical method yield essentially the same aromaticity values within experimental error for each of these four samples. Methods 1 and 2 appear to give values beyond the experimental error in certain instances, thus indicating that more reliable results are obtainable from Method 3, carbon-13, and the graphical method. Of the latter methods, the graphical method yields an answer more quickly since in the carbon-13 method, data accumulation requires approximately eight hours of time and Method 3 requires a complete analysis of the spectrum. Therefore, the graphical method is the best choice for a rapid and reliable determination of the aromaticity for these samples. Generalizing this conclusion to other samples may not be warranted, however. The graphical method works best here because the particular values of x and y assumed in formulating the graph appear to be valid. If good estimates of the values of the parameters x and y can be determined, Method 1 may be the most appropriate way to estimate the carbon aromaticity.

If average parameters of the aromatic fraction of the petroleum fraction are desired, then a choice between Method 2, Method 3, and the carbon-13 method must be made since Method 1 and the graphical method yield no additional average parameters. Table V-6 lists a number of average parameters calculated by the three methods for the same four samples. The molecular weights listed for Method 2 and carbon-13 are those determined by mass spec-

Table V-5. Aromaticities Determined for FCC Charge Stocks by Nuclear
Magnetic Resonance

Sample Method[a]	F002	F004	F014	F016
^{1}H No. 1	0.34	0.34	0.29	0.43
^{1}H No. 2	0.39	0.36	0.33	0.48
^{1}H No. 3	0.38	0.30	0.35	0.41
^{13}C	0.37	0.31	0.35	0.38
Graphical	0.37	0.29	0.34	0.38

[a]For a description of the methods, see the text.

Table V-6. Average Parameters Calculated from Proton and Carbon-13
Spectra of Aromatic Fractions of FCC Charge Stocks

Sample	F014			F002		
Parameters[a]	Method 2	Method 3	^{13}C Method	Method 2	Method 3	^{13}C Method
MW	312.6	285.3	312.6	296.4	272.2	296.4
n	4.21	3.92	4.20	3.64	3.62	3.64
f	5.63	5.56	5.49	5.50	5.49	5.73
%AS	51.50	52.63	50.77	50.75	50.65	51.75
#C_A	6.98	7.28	7.33	8.57	7.60	7.98
#C_1	6.59	6.64	6.52	7.14	6.80	7.29
R_A	1.19	1.32	1.41	1.71	1.40	1.35
R_S	3.39	3.49	3.31	3.63	3.44	3.77
R_N	0.76	0.67	0.39	0.62	0.56	1.23
f_a	0.33	0.35	0.35	0.39	0.38	0.37
#C_T	23.0	21.0	--	22.0	20.1	--
#H_T	36.2	33.0	--	32.4	30.9	--
%Monoaro-matics	--	68	--	--	60	--

109

Table V-6 (continued)

Sample Parameters[a]	F004			F016		
	Method 2	Method 3	^{13}C Method	Method 2	Method 3	^{13}C Method
MW	319.5	359.2	319.5	318.3	252.0	318.3
n	4.97	4.72	4.97	3.35	3.19	3.35
f	5.72	5.55	6.19	5.49	5.48	6.48
%AS	56.37	56.61	58.30	50.28	51.33	54.61
$\#C_A$	8.36	7.82	7.17	11.33	7.55	9.06
$\#C_1$	5.27	6.91	5.51	7.43	6.78	8.07
R_A	2.55	1.45	1.83	2.95	1.39	1.50
R_S	2.97	3.91	3.21	3.74	3.48	4.41
R_N	0.77	0.47	2.10	0.71	0.68	3.29
f_a	0.36	0.30	0.31	0.48	0.41	0.38
$\#C_T$	23.8	26.3	--	23.9	18.6	--
$\#H_T$	34.0	43.2	--	30.9	27.8	--
%Monoaromatic	--	55	--	--	61	--

[a]See Table V-1 for definitions.

tral carbon number distributions, whereas the molecular weights under Method 3 are calculated in the analysis by Method 3.

Close examination of Table V-6 reveals that in many instances all three methods yield essentially the same value for the average parameters although different methods are used. Some of the parameters which agree well are n, %AS, and f_a. The other parameters which are shown tend to have discrepancies between the different methods. The question which arises now is which one of the methods gives the most accurate description of the sample. First let us turn to the three parameters $\#C_A$, $\#C_1$, and R_A. R_A is directly related to $\#C_A$ and $\#C_1$ in Method 2 and carbon-13 method. In samples F004 and F016, the value of R_A for Method 2 is inconsistent with $\#C_A$. For instance, F014 and F016 are shown to have an average number of 2.55 and 2.95 rings, respectively, and only 8.36 and 11.33 total carbons in the aromatic ring system. Thus, these two parameters are not consistent with each other, because R_A predicts a much larger $\#C_A$ than is actually determined. Likewise, there is a discrepancy in the carbon-13 method for these two parameters although not as serious as Method 2. An analysis of equations 5, 6, 8,9 and 11 indicates that $\#C_A$, $\#C_1$, and R_A are directly dependent upon f, H, and the molecular weight in a multiplicative manner in Method 2. Thus, if f has not been estimated properly or the hydrogen weight per cent (H) is in error or an incorrect molecular weight has been used, a true description of the aromatic composition of the sample is not given. Such a propagation of errors is a shortcoming present in most average parameter calculations. The magnitude of these errors has been examined by Shenkin (1983). Little propagation of error is observed in the Method 3 analysis, however, since it involves only spectral information.

From this comparison and investigation of the sources of inconsistencies observed within the various methods of calculating the average parameters, the end result is that all of these methods give approximately the same average parameters for these samples provided accurate elemental analysis and average molecular weights are available for input data in Method 2 and the carbon-13 method. The most economical and straightforward method is Method 3 since only a proton spectrum is required to obtain the average parameters.

These methods for calculating average molecular parameters of a petroleum fraction have all been developed for application to aromatic fractions of samples. The extension of these methods to samples containing both the saturate and aromatic fractions does not yield reliable information except for the aromaticity. If additional assumptions are made and the ratio of aromatic and saturate fractions is known, then somewhat reliable parameters can be determined; however, in order to determine the amount of saturates and aromatics present, separation of the two must be performed and thus the aromatic fraction can be analyzed in detail.

Similar methods for calculating average parameters have been developed for coal liquids by Bartle et al. (1975) and others. Different average parameter methods are required for coal derived materials since large aromatic rings and molecules containing more than one aromatic ring system are frequently present. An assessment similar to the assessment of petroleum methods can be made for these structural characterization methods.

REFERENCES

Bartle, K. D., Martin, T. G. and Williams, D. F., Fuel, 54, 226 (1975).

Brown, J. K., and Ladner, W. R., Fuel, 39, 87 (1960).

Clutter, D. R., Petrakis, L., Stenger, R. L., Jr., and Jensen, R. K., Anal. Chem., 44, 1395 (1972).

Grant, D. M., and Paul, E. G., J. Amer. Chem. Soc., 86, 1984 (1964).

Hirsch, E., and Altgelt, K. H. Anal. Chem., 42, 1330 (1970).

Jewell, D. M., Ruberto, R. G. and Davis, B. E., Amer. Chem. Soc. Div. Petr. Chem., 17(1), A55 (1972a).

Jewell, D. M., Ruberto, R. G. and Davis, B. E., Anal. Chem., 44, 2318 (1972b).

Knight, S. A., Chem. Ind., 1920 (1967).

Retcofsky, A. L., and Friedel, R. A., "Spectrometry of Fuels", R. A. Friedel, Ed., Plenum Press, New York (1970), Chapter 8.

Shenkin, P. S., Prepr. Amer. Chem. Soc. Div. Petr. Chem., 28(5), 1367 (1983).

Williams, R. B., "Symposium on Composition of Petroleum Oils, Determination and Evaluation", ASTM Spec. Tech. Publ., 224, 168-94 (1958).

Chapter VI

AVERAGE MOLECULE CONSTRUCTION

The concept of an average molecular structure has been used frequently in the interpretation and presentation of NMR spectra of liquid fuels. Essentially, the goal of this type of characterization is to generate one or several molecular structures that have the same distribution of functionalities as the fuel. The procedure is best described through a detailed example.

Suppose that we wanted to develop an average molecular structure using the data in Table VI-1. The molecular weight and the elemental analysis data are combined to yield the numbers of carbon and hydrogen atoms in the average molecule. The proton and carbon NMR spectra are then used to determine the numbers of aromatic carbon, aliphatic carbon, aromatic hydrogen, alpha hydrogen, beta hydrogen, gamma hydrogen, oxygen, nitrogen and sulphur in an average molecule. This results in a detailed molecular formula, as shown in Table VI-2. The problem now becomes one of matching a structure to this molecular formula.

Oka (1977) has presented a systematic way of generating molecular structures from the detailed molecular formula. In this approach, the atomic concentrations in the molecular formula of Table VI-2 are rounded to the nearest whole integer.   A computer program is then used to generate all combinations of functional groups that result in the correct molecular formula. Thirty four different functional groups were used in the Computer Assisted Molecular Structure Construction (CAMSC), as originally described by Oka (1977).   These groups are shown in Figures VI-1 and VI-2. Later, a revised version of the procedure incorporated two improvements into the program (Chang, 1982). First, nitrogen functional groups were added and second, a method was developed for assessing the likelihood of various structures generated by the program.   Priorities were established for the structures based on the assumed likelihood of various ring sizes, degrees of substitution, and numbers of ring clusters, quaternary carbons and tertiary carbons. The structure of the computer program is presented in the flow chart of Figure VI-3 and the nomenclature is defined in Table VI-3.

If this algorithm is applied to the data of Table VI-2, the structure shown in Figure VI-4 results.   It is instructive at this point to consider the strengths and weaknesses of using average structures to represent complex fuel fractions.   One of the primary strengths of the method is that an average

Table VI-1

Average Molecule Construction: Analytical Data for a
Supercritical-gas Extract of Coal*

Elemental Analysis

C	88.4%
H	8.4%
O	2.1%
Nonphenolic O	1.4%

^{1}H NMR Data

Aromatic H (δ > 6.0 ppm)	28.7%
Methylene bridge H (4.3 > δ > 3.2 ppm)	2.9%
Other alpha H (3.2 > δ > 1.8 ppm)	32.2%
Beta H (1.8 > δ > 1.1 ppm)	25.1%
Gamma methyl H (1.1 ppm > δ)	11.1%

Average Molecular Weight = 310

^{13}C NMR Data

Aromatic C	66%
Aliphatic C	34%

*From Bartle et al., 1975.

Table VI-2

Average Molecule Construction: Average Molecular Formula for a

Supercritical-gas Extract of Coal*

Atomic Type	Number of Atoms in an Average Molecule	Input Data for CAMSC
Aromatic C	16.1	16
Aliphatic C	7.5	7
Aromatic H	7.5	8
Alpha H	8.4	9
Methylene H	0.8	0
Beta H	6.6	9
Gamma H	2.9	
Phenolic O		0
Etheric O		0

*From Bartle et al., 1975.

The oxygen atoms have been replaced with an equivalent number of aromatic carbons.

Table VI-3

Notation for CAMSC Program

C_{ar}	number of aromatic carbons
i	aromatic groups
n_i	number of aromatic cluster i
NC_i	number of aromatic carbons in aromatic cluster i
H_{α}	number of α-hydrogens
H_{β}	number of β-hydrogens
H_{ar}	number of aromatic hydrogens
SS	number of aromatic substituted sites
AS_i	number of aromatic sites in aromatic cluster i
j	aliphatic groups
HA_j	number of H increased or decreased by addition of parameter j
S_j	number of sites occupied by parameter j
H_j	number of α-hydrogens increased or decreased by addition of parameter j
C_j	number of carbons in parameter j
k_j	number of aliphatic parameter j
H_{al}	number of aliphatic hydrogen
C_{al}	number of aliphatic carbon

Group Number	Aromatic Groups	NC_i	AS_i
1	C = C	2	4
2		6	6
3	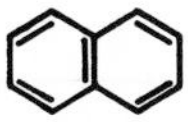	10	8
4	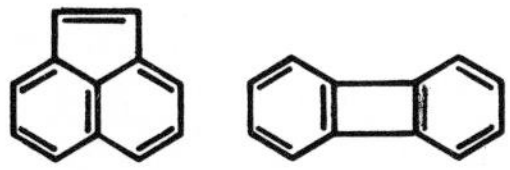	12	8
5	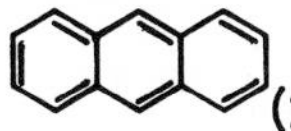 $(2)^{\alpha}$	14	10
6	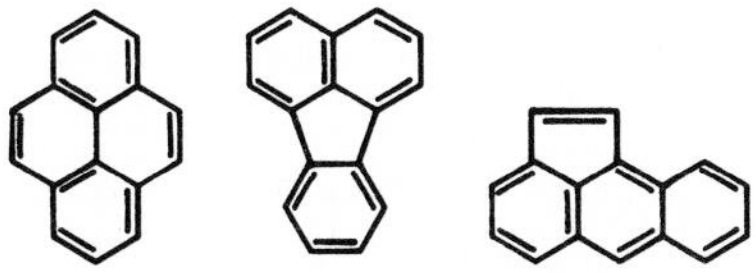	16	10
7	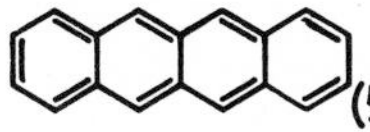 (5) 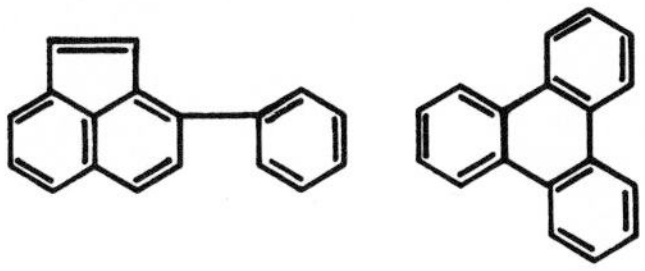	18	12
8	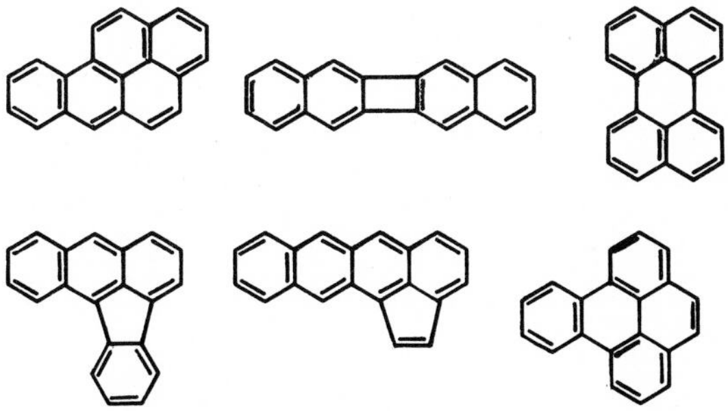	20	12

Figure VI-1.   Aromatic Groups for Computer Assisted Average Molecule Construction (CAMSC)

118

Figure VI-1 (continued)

Group Number	Aromatic Groups	NC$_i$	AS$_i$
9	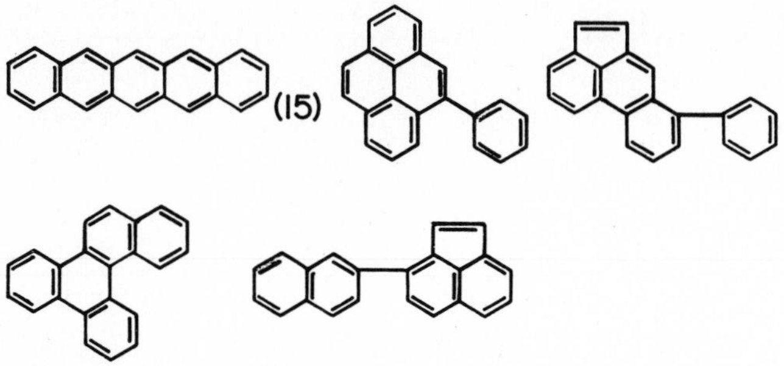 (15)	22	14
10	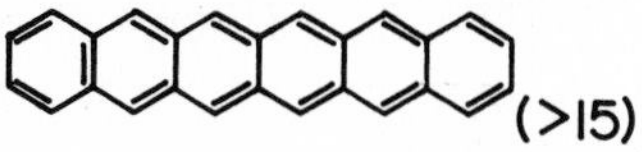 (>15)	26	16

Figure VI-2. Aliphatic Groups for Computer Assisted Average Molecule Construction (CAMSC)

Group Number	Aliphatic Groups	HA_j	S_j	H_j	C_j
11	$-CH_2-$	2	2	2	1
12	$-CH_2CH_2-$	4	2	4	2
13	$-\!\!-$	0	2	0	0
14	$-CH_3$	3	1	3	1
15	$-CH_2CH_3$	2	1	5	2
16	$-CH(CH_3)_2$	1	1	7	3
17	$\odot-CH_2$	-1	0	2	1
18	$\circ-CH_2$	0	0	2	1
19	$-\!\!-\odot$	-1	1	-1	0
20	$-\!\!-\circ$	0	1	-1	0
21	$\odot-\!\!-\odot$	-2	0	-2	0
22	$\odot-\!\!-\circ$	-1	0	-2	0
23	$\circ-\!\!-\circ$	0	0	-2	0
24	 (ring structure)		2	4	2
25	(ring structure)	$4(5\ or\ 6)^a$	2	6	3
26	 (ring structure)	4(6)	2	8	4
27	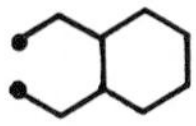 (ring structure)	4(5)	2	14	8
28	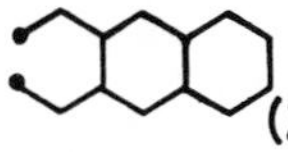 (ring structure) $_{(2)}{}^b$	4(5)	2	20	12

Figure VI-2 (continued)

Group Number	Aliphatic Groups	HA_j	S_j	H_j	C_j
29	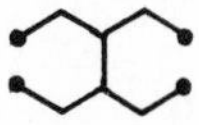	8(9)	4	10	6
30	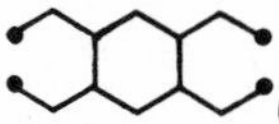 (2)	8(10)	4	16	10
31	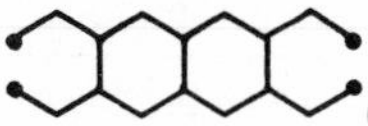 (5)	8(10)	4	22	14
32	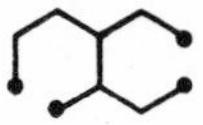	7(9)	4	10	6
33	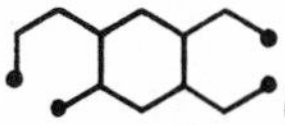 (3)	7(11)	4	16	10
34	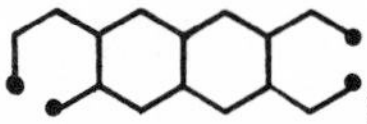 (16)	7(11)	4	22	14

Key: • aromatic carbons, ⊙ C_α carbons, ○ C_β carbons

[a] Estimated correct values for alicyclic groups since the H_β hydrogens in the alicyclic groups become H_α hydrogens in some degree, as far as the [1]H n.m.r. chemical shift is concerned

[b] Number of possible arrangements (isomers)

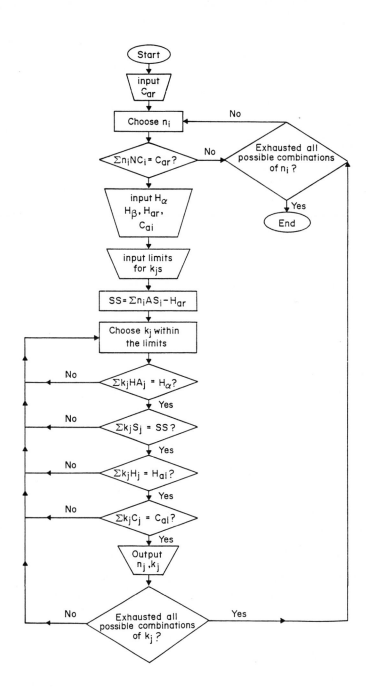

Figure VI-3.   Flow Diagram for CAMSC Program

122

Structure obtained by:

CAMSC

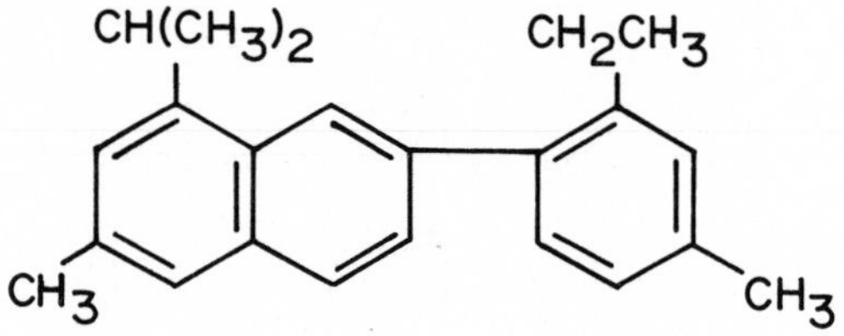

Bartle

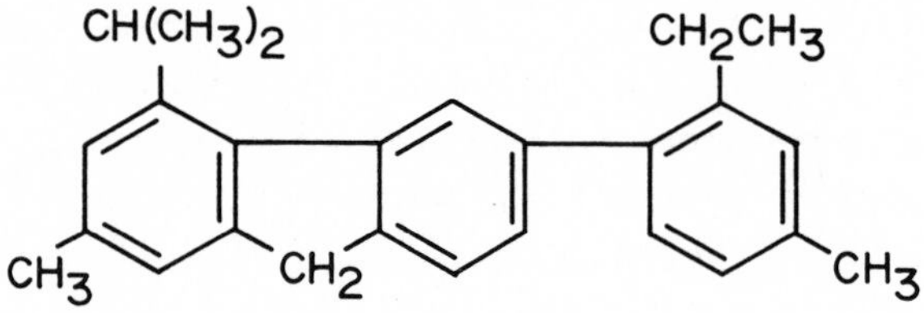

Figure VI-4. Average Molecular Structures

structure gives a clear picture of the types of structures that may be present in a mixture. A second strength of the method is that it provides a reasonable mechanism for interfacing analytical data with property estimation methods. Property estimation is performed by assuming that the properties of the mixture will be the same as the properties of the average structure. The properties of the average structure are obtained from either pure component data or group contribution methods. This topic will be addressed in more detail in Chapter VIII.

Counterbalancing these advantages are two severe disadvantages. First, the analytical data, on which an average structure is based, are averaged over all of the components in the mixture. The average structure is unlikely to be the dominant component in the mixture and may not even be present. A second disadvantage of average molecule construction is an outgrowth of allowing only integer numbers of atoms in the average structure. Consider the situation when the average molecular formula yields values of 0.33 mole of OH groups per molecule and 0.2 mole of S per molecule. One in three molecules will contain a phenolic group. One in five will contain a sulfur group. Approximately one in fifteen will contain both an oxygen and a sulfur group. These heteroatoms will profoundly effect the properties of the mixture, but it is impossible to accurately represent their concentration with just one or two average structures. On the other hand, if 15 structures were chosen to represent a mixture it would probably be better to identify the 15 most dominant components in the mixture. An additional problem associated with allowing only integer numbers of groups is the question of the stability of the calculations. In the example outlined in Tables VI-1 and VI-2, we could change the value of the number of aliphatic carbons from 7.5 to 7.7. The structure generated with an average molecular formula containing 8 aliphatic carbons can be vastly different than the structure generated with 7 aliphatic carbons. Thus, the average structures are extremely sensitive to small perturbations in the data.

The remedy for these disadvantages is to allow non-integer values of various functionalities in the average structure. This topic will be discussed in detail in Chapter VIII, but first we will present a general method for estimating non-integer functional group concentrations in complex mixtures.

124

References

Bartle, K. D., Martin, T. G. and Williams, D. F., Fuel, $\underline{54}$, 226 (1975)

Chang, P., Oka, M. and Hsia, Y. P., Energy Sources, $\underline{6}$, 67 (1982).

Oka, M., Chang, M. and Gavalas, G. R., Fuel, $\underline{56}$, 3 (1977).

Chapter VII

FUNCTIONAL GROUP ANALYSIS

VII.A METHODOLOGY FOR FUNCTIONAL GROUP ANALYSIS

One of the three basic methods developed for the characterization of complex mixtures, such as the various liquid fuels, is known as functional group
analysis (Petrakis, 1983a; Allen, 1985). While the methods discussed in
Chapters V and VI rely almost exclusively on NMR data, this method addresses
the problem of how to combine NMR data with other sources of data into a single structural characterization. The goal of functional group analysis is to
determine a set of functional group concentrations that characterize the
structure of a complex mixture. The rationale for using functional group concentrations to characterize liquid fuels is based on three considerations
(Allen, 1984).

1. While the number of molecular species present in liquid fuels is gen-
    erally large, the number of different functionalities is relatively
    small. Therefore, when a limited amount of data is available, estimat-
    ing functional group concentrations is a more attractive alternative
    than estimating molecular concentrations.

2. Detailed characterizations of complex mixtures such as liquid fuels
    should be based on several sources of analytical data. The integration
    of many sources of data into a single structural characterization
    requires that all of the data be converted to a single basis. A concen-
    tration basis is preferred since some sources of data, such as NMR, do
    not provide information on the molecular species present, but only on
    the relative concentrations of atoms and functionalities.

3. The concentrations of functional groups can be used to estimate thermo-
    dynamic and physical properties such as heats of formation per unit
    mass, heat capacities (Le and Allen, 1985) and critical properties
    (Lydersen, 1955). Estimates of thermodynamic and physical properties
    based on functional groups are more easily calculated and may be more
    accurate than estimates based on either average molecules or average
    structural parameters.

The first step in determining the desired group concentrations is to pro-
pose a set of functional groups representing the possible structures present
in the fuel. The choice of the set of functional groups is one of the most

qualitative aspects of this method of characterization. Factor analysis can be used to select appropriate groups, but in general the choice of groups is based on experience or on qualitative experimental data provided by IR and mass spectrometry. A typical set of groups is shown in Figure VII-1. The set must be adequate to account for the observed data, yet it must be as concise as possible since the number of meaningful concentrations that can be determined is limited by the available data.

Once the functional groups have been specified, concentrations which satisfy the available analytical data must be found. This is done by relating the concentrations of the functional groups to the experimental data through a set of balance equations. As an example of how these balance equations are formulated, consider the balance equation for α hydrogen using the groups in Figure VII-1. The concentration of α hydrogen in the sample is calculated by using ^{1}H NMR and elemental analysis data. This concentration must equal the concentration of each functional group times the number of α hydrogens in that group, summed over all groups. For the set of groups shown in Figure VII-1 the balance equation for α hydrogen takes the following form.

$$3y_8 - y_9 + 2y_{11} + {}^4y_{12} + {}^4y_{13} + 3y_{14} + 6y_{15} = b_3 \tag{1}$$

where the y_j are the functional group concentrations and b_3 is the concentration of alpha hydrogen. The set of all balance equations can be expressed in the matrix form

$$\sum_{j=1}^{n} A_{ij}y_j = b_i \quad (i=1,\ldots,m) \tag{2}$$

where the y_j (j=1,...,n) represent the unknown functional group concentrations, b_i (i=1,...,m) are quantities representing the elemental and NMR data, and the A_{ij} represent stoichiometric coefficients. If the functional groups of Figure VII-1 are combined with the data of Table VII-1, the stoichiometric coefficients and data vector b of Tables VII-1 and VII-2 result.

In addition to the constraints imposed by the data, we know that the concentrations cannot be negative.

$$y_j \geq 0 \tag{3}$$

For the mathematical problem presented by equations 2 and 3, two general classes of solution are possible. In the first class, the number of balance equations (2) is greater than the number of unknown functional group concentrations. In this case, the concentrations are determined using weighted least squares methods. Heavy oils and other fuels which yield detailed ^{13}C spectra frequently fall into this class.

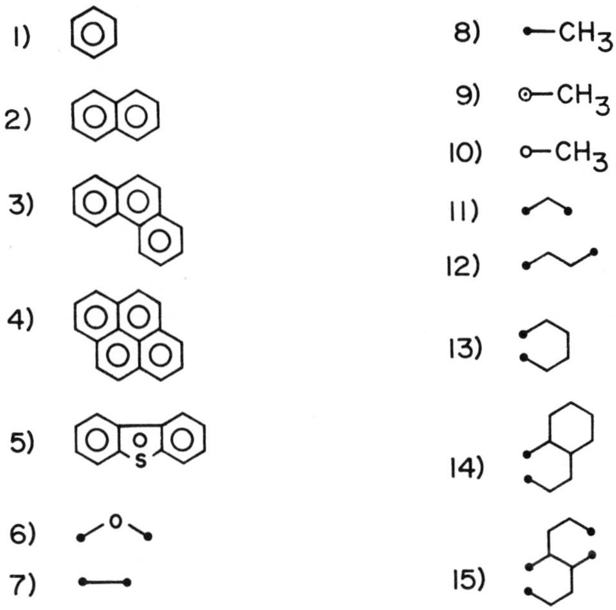

Notation:

— bound directly to an aromatic ring

∘— bound to a carbon alpha to an aromatic ring

∘— bound to a carbon beta or further from an aromatic ring

Examples:

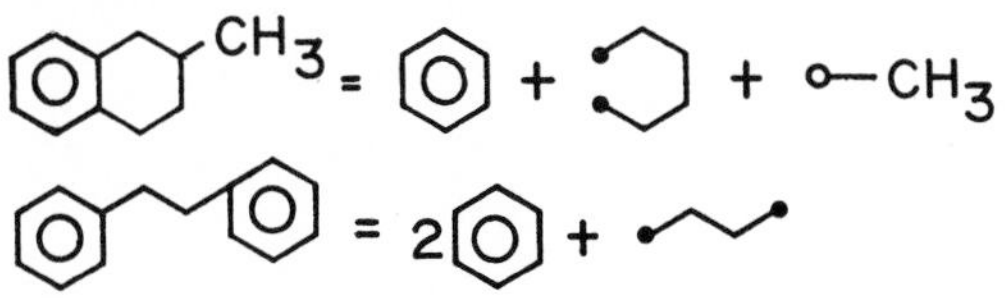

Figure VII-1. Functional Groups for a Neutral Fraction of a Light Coal Liquid Distillate

Table VII-1

Analytical Data for a Neutral Fraction of a Light Coal Liquid Distillate

Elemental Composition Data

%C	%H	%O	%S
91.4	7.3	0.29	0.62

^{1}H NMR Data

% aromatic (9.0-5.0 ppm)	% alpha (5.0-1.9 ppm)	% beta (1.9-1.0 ppm)	% gamma (1.0-0.5 ppm)
43.8	33.0	19.3	4.0

Concentration Data, b_i (moles/100 g)

b_1 carbon	b_2 aromatic hydrogen	b_3 alpha hydrogen	b_4 beta hydrogen	b_5 gamma hydrogen	b_6 oxygen	b_7 sulfur
7.62	3.18	2.39	1.40	0.29	0.018	0.020

Table VII-2

Stoichiometric Coefficients, A_{ij}, relating constraint i to functional group j

Functional group j (a)

constraint i (b)	1	2	3	4	5	6	7	8	9	10	11	12	13	14	15
1. carbon	6	10	14	16	12	0	0	1	1	1	1	2	4	8	6
2. aromatic hydrogen	6	8	10	10	8	-2	-2	-1	0	0	-2	-2	-2	-2	-4
3. α hydrogen	0	0	0	0	0	0	0	3	-1	0	2	4	4	3	6
4. β hydrogen	0	0	0	0	0	0	0	0	3	-1	0	0	4	11	4
5. γ hydrogen	0	0	0	0	0	0	0	0	0	3	0	0	0	0	0
6. oxygen	0	0	0	0	0	1	0	0	0	0	0	0	0	0	0
7. sulfur	0	0	0	0	1	0	0	0	0	0	0	0	0	0	0

(a) Functional group numbers are defined in Figure VII-1.

(b) Constraint vectors are given in Table VII-1.

In the second class of solutions, the number of equations is smaller than the number of unknowns. In this case, equations 1 and 2 can yield either no solutions or an infinite number of solutions. If the equations have no solution, then the proposed set of functional groups is insufficient to describe the observed data and must be revised. If a space of solutions exist, then the mixture can be characterized by selecting a single solution from the feasible space. Selecting a single solution provides a valid structural characterization because the range of feasible functional group concentrations is limited, even though the number of solutions is mathematically defined as infinite.

To select a single solution from the feasible region, we use the following computational procedure. The concentrations $y_1,...,y_n$ are chosen such that a function P $(y_1,...,y_n)$ is minimized, subject to the constraints of equations 2 and 3. The form of the function P depends on what, if any, data are available in addition to elemental analysis and NMR. Data other than elemental analysis and NMR can be introduced as additional balance equations or can be incorporated into the function P. The method for handling the data is chosen on the basis of the accuracy and precision of the data. Highly precise and accurate data are used in the balance equations. More qualitative data are utilized in formulating the minimization functions.

Three different minimization functions will be discussed. The first is a general function which is applied when all of the available data have been used in formulating the balance equations. The second and third functions are designed to utilize high- and low-resolution mass spectrometric data, respectively. Other types of minimization functions are possible, and may be formulated to take advantage of additional data sources. For comparison, all three methods are applied to a set of coal liquids.

VII.A.1 General Minimization Function

The general minimization function is used if no information other than that used in the balance equations is available on the relative abundance of the various functional groups. We formulate the function by assuming that all functional groups are equally probable. In principle, this is similar to an entropy maximization. As a concrete example of this, consider the distribution of carbons among groups 1-5 (aromatic) and 8-15 (aliphatic) of Figure VII-1. If these 13 groups were equally probable, each would contain 1/13 of the carbon. This statement can be implemented by minimizing

$$P_1 = \sum_{A_{1j} \neq 0} \left(y_j - \frac{b_1}{13A_{1j}} \right)^2$$

(4)

subject to equations 2 and 3.

However, there is no reason to single out carbon to equally distribute among functional groups. For aromatic hydrogen, the function to be minimized is

$$P_2 = \sum_{A_{2j} \neq 0} \left[\frac{1}{5A_{2j}} (b_2 + 2y_6 + 2y_7 + y_8 + 2y_{11} + 2y_{12} + 2y_{13} + 2y_{14} + 4y_{15}) - y_j \right]^2 \quad (5)$$

wherein the total number of aromatic hydrogen positions to be distributed among the five functional groups is equal to the number of aromatic hydrogens plus the number of aromatic substituents. Similarly, the expressions for the distribution of the remaining experimental quantities of α, β, and γ hydrogen, oxygen and sulfur concentrations, are

$$P_3 = \sum_{A_{3j} \neq 0} \left[\frac{1}{6A_{3j}} (b_3 + y_9) - y_j \right]^2 \quad (6)$$

$$P_4 = \sum_{A_{4j} \neq 0} \left[\frac{1}{4A_{4j}} (b_4 + y_{10}) - y_j \right]^2 \quad (7)$$

$$P_5 = \left(\frac{b_5}{3} - y_{10} \right)^2 \quad (8)$$

$$P_6 = (b_6 - y_6)^2 \quad (9)$$

$$P_7 = (b_7 - y_5)^2 \quad (10)$$

minimizing any of the functions P_i (i=1,...,7) would express a lack of preference among functional groups with respect to a particular concentration, e.g., aromatic hydrogen. If we minimize the sum of the individual P_is we will get solutions that express the lack of preference among functional groups with respect to all of the concentrations. Thus, P_8 is the function to be minimized when no data other than NMR and elemental analysis are available.

$$P_8 = \sum_{i=1}^{7} P_i \quad (11)$$

VII.A.2 Minimization Function Based on High Resolution Mass Spectral Data

Equation 11 is utilized if no data is available other than that used in formulating the balance equations. If high resolution mass spectra are available, however, the situation is quite different. High-resolution mass spectra provide a wealth of information on the structure of fossil fuel fractions.

For typical coal liquids, approximately 90 separate peaks can be resolved in the spectra. Since high-resolution mass spectrometry can resolve masses to one part in 150,000, heteroatoms can be resolved from the hydrocarbon background, and precise empirical formulae can be assigned to each peak in the spectra. Quantitative interpretation of the spectra is difficult, however, because the relationship between peak intensities and concentrations is not always known and peaks are often difficult to resolve. Thus, the mass spectra are less quantitative than the NMR and elemental analysis data are more appropriately used in the minimization function than in the balance equations.

In constructing the minimization function, functional group concentrations are determined from the mass spectra. This is done by assigning a structure to each molecular formula appearing as a peak in the mass spectrum. The types of structures assigned to the peaks are given by Herod (1981). Each of the structures is broken into its constituent functional groups. The concentration associated with each mass peak is multiplied by the numbers of groups associated with the peak. The apparent concentrations of a set of functional groups are then calculated by summing over all mass peaks.

This method of interpreting the high-resolution mass spectra involves several approximations. First, it is assumed that each peak in the mass spectrum can be represented by one structure. Many of the molecular formulas have more than one isomer. In some cases, this is not a concern since the different isomers can be constructed from identical functional groups. When isomers are composed of different functional groups, the structure selected to represent the peak is the structure considered most likely to be found in the liquid being considered.

A second assumption is that a concise set of functional groups can be used to represent all of the structures assigned to the peaks. The functional groups are chosen to provide as complete a description as possible of the structure of the fuels, while maintaining a low number of functional groups. Groups needed to construct the structures assigned to the peaks but having extremely low concentrations are incorporated into the concentrations of similar functional groups. This assumption does not present a serious problem since the vast majority of the structures can be constructed exactly from a small set of functional groups.

The set of functional group concentrations which most closely resembles the concentrations derived from the high-resolution mass spectra, yet still satisfies the elemental composition and NMR data, is determined by minimizing equation 12 subject to constraints 2 and 3

$$P = \sum_{j=1}^{n} (y_j - f_j)^2 \qquad (12)$$

where f_j is the concentration of functional group j determined from the mass spectral data. Equation 12 may be thought of as the minimum distance between the point defined by the mass spectral concentrations and the space of feasible solutions defined by equations 2 and 3. The minimum value of P serves as a measure of the agreement between the concentrations determined from the high-resolution mass spectral data and the rest of the analytical data.

VII.A.3 Minimization Function Based on Low Resolution Mass Spectral Data

The approximations involved in interpreting the high-resolution mass spectra also apply to the low-resolution spectra, so the data are treated in an analogous manner. Structures are assigned to the peaks resolved in the low-resolution spectrum and estimates of the functional group concentrations are calculated.

Since any alkyl chains present in the structures are removed during the high-voltage ionization used in obtaining the low-resolution spectrum, alkyl chains are not detected. Thus, the low-resolution spectra only give data on the relative concentrations of aromatic and hydroaromatic groups. Functional group analysis is used to find the set of concentrations which most closely resemble the concentration ratios predicted by the low-resolution mass spectra. Equation 13 is the minimization function appropriate for this purpose and the minimum of equation 13 serves as a measure of the agreement between the mass spectral concentrations and the NMR and elemental analysis data.

$$P = \sum_{j=1}^{n} (y_j/y_1 - f_j/f_1)^2 \qquad (13)$$

The method of functional group analysis has been applied to mixtures of known composition and has been demonstrated in detail on a variety of fuel products. A hydrodesulfurized coal liquid will be used to present typical results for fuel products.

The mixtures of known composition are described in Table VII-3. Two sets of calculations were made on the mixtures. First, functional group concentrations were estimated for both mixtures using only elemental analysis and NMR data. These concentrations indicate how well the general method estimates absolute concentrations and changes in concentrations between samples. Next, concentrations for one of the samples were estimated by obtaining a high-resolution mass spectrum of the mixture. Concentrations were estimated directly from the mass spectrum and by using equation 12 as the minimization function in the functional group analysis algorithm. The results in Table

Table VII-3
Functional Group Analysis of Model Compound Mixtures

FUNCTIONAL GROUP ANALYSIS OF MODEL COMPOUND MIXTURES

Mixture Composition

	Mixture #1 (g)	Mixture #2 (g)		Mixture #1 (g)	Mixture #2 (g)
	13.2	13.2		13.1	13.1
	3.3	3.3		31.4	14.3
	14.0	14.0		6.7	6.7
	7.4	7.4		7.8	7.8
	6.9	6.9		6.3	6.3
	7.0	7.0			

VII-4 show that the general method, utilizing only elemental analysis and NMR data, gives fair estimates of absolute concentrations and good estimates of changes in concentrations. When additional data, particularly high-resolution mass spectra, are available, the estimates become more accurate. While the results for coal liquids may not be as accurate as for these simple mixtures, these results give an estimate of the uncertainties involved.

We now proceed to the examination of a fuel derived example, a hydrodesulfurized coal liquid. The method for obtaining the hydrodesulfurized coal liquid sample is described by Petrakis (1983a).

The set of 15 functional groups adopted for this samples is listed in Figure VII-1. This list is not meant to be an exhaustive compilation of all functional groups present in the coal liquid. Instead, it contains groups representative of the basic structural features of the liquid, i.e., aromatic clusters, hydroaromatic groups, aliphatic chains, bridges, and oxygen-containing groups. Nitrogen compounds and phenolic structures were not considered because the procedure used to obtain the liquid removes nitrogen and phenolic functionalities. Additional groups, such as larger condensed aromatic structures and longer alkyl chains, may be necessary for some samples of coal liquids but were not needed to explain the analytical data presented here.

Having specified the functional groups, the balance equations can be immediately written down. For example, the balance for α hydrogen is

$$3y_8 - y_9 + 2y_{11} + 4y_{12} + 4y_{13} + 3y_{14} + 6y_{15} = b_3 \tag{14}$$

where b_3 is the experimentally determined concentration of α hydrogen, y_j is the concentration of functional group j (as defined in Figure VII-1), and the numerical coefficients are the stoichiometric A_{ij}.

There are seven balance equations in the 15 unknowns $y_1, \ldots, y_{15}$. Table VII-1 gives the concentrations obtained from the elemental analyses and the ^{1}H NMR integrals for the four distinct hydrogen environments and for total carbon, oxygen and sulfur. Table VII-2 gives the stoichiometric coefficients which relate each of the 15 functional groups to the same hydrogen molecular environments and C, O and S elemental determinations.

Functional group concentrations were determined by minimizing the P function defined in equation 11, subject to equations 2 and 3. This procedure requires only elemental analysis and ^{1}H NMR data. The algorithm of Luus and Jaakola (1973) was used in the minimization. The results are listed in Table VII-5. Equations 12 and 13 were also minimized subject to equations 2 and 3. This procedure determined the concentrations which most closely resemble the concentrations derived from the low- and high-resolution mass spectral data.

Table VII-4

Functional Group[a]	Mixture No. 1		Mixture No. 2			
	True Concentration (moles/100 g)	Concentration Estimated Using Eqs. 2,3,11	True Concentration (moles/100 g)	Concentration Estimated Using Eqs. 2,3,11	Concentration Estimated Using Mass Spectrum	Concentration Estimated Using Eqs. 2,3,12
1. Monoaromatics	0.585	0.463	0.585	0.420	0.503	0.516
2. Diaromatics	0.309	0.279	0.199	0.245	0.211	0.215
3. Triaromatics	0.075	0.148	0.075	0.112	0.098	0.093
7. Biphenyl bridge	0.129	0.000	0.129	0.000	0.113	0.096
8. Alpha methyl	0.251	0.250	0.141	0.140	0.177	0.140
9. Beta methyl	0.202	0.240	0.092	0.150	0.108	0.098
11. Methylene bridge	0.143	0.153	0.143	0.135	0.097	0.106
12. Ethylene bridge	0.076	0.109	0.076	0.138	0.063	0.099
13. Hydroaromatic ring	0.165	0.137	0.165	0.122	0.121	0.162

a)Functional group structures are given in Figure VII-1.

The results are listed in Table VII-5. The concentrations derived from the mass spectra are given in Table VII-6.

The results obtained by minimizing equations 11, 12, and 13 are quite similar. The only significant differences occur in functional groups 1, 7, 11 and 12. The various results disagree on the distribution of aliphatic hydrogen among the various types of alkyl groups and the importance of biphenyl bridges. The only independent experimental evidence available on these concentrations for the hydrodesulfurized samples is estimates of functional group 11 concentration obtained from ^{1}H NMR. Assuming all hydrogens with chemical shifts between 3.4 and 5.0 ppm are in group 11 structures yields concentrations of 0.07-0.10 mol/100 g. This is in reasonable agreement with the results based on the mass spectra. ^{13}C NMR spectra could be used to better define the distribution between alkyl chain and bridge groups. Overall, the functional group concentrations obtained by minimizing P functions derived using the three approaches are in reasonable agreement.

The concentrations derived directly from the mass spectra, shown in Table VII-6, very nearly match those derived from the functional group analysis. Considering the approximate nature of the NMR band assignments and the assumptions involved in assigning the mass spectra structures, the agreement is quite good. The close agreement of functional group concentrations based on two independent sources of data lends support to the validity of functional group characterization of mixtures.

Finally, the results of the functional group analysis method were compared to the results of an established method of structural analysis. The average structural parameters defined by Clutter et al. (1972) were used for the comparison and the results are shown in Table VII-7. The average parameters were calculated in the standard manner by using ^{1}H NMR spectra and by direct evaluation from the functional group concentrations. The agreement between the two methods is excellent, indicating that functional group distributions provide at least as much information as average parameters.

This section has outlined the general procedure for estimating functional group concentrations. In the next sections we will demonstrate the method's scope by applying it to coal liquids, shale oils, oil derived from tar sands and heavy oils. Functional group analysis is a novel and powerful technique, as the following concrete examples illustrate.

VII.B APPLICATION TO COAL LIQUIDS

In the previous section we described a method for estimating the concentrations of functional groups in liquid fuels from NMR and other data. In this section we consider the application of this method to coal liquids. In subsequent sections, heavy oils, shale oils and fuels derived from tar sands will

Table VII-5

Functional Group Concentrations in a Light Coal Distillate (in moles/100 g)

Functional Group[a]	General Results (Eqs. 2,3,11)	Results	
		Using High Resolution MS (Eqs. 2,3,12)	Using Low Resolution MS Eqs. 2,3,13)
1. Monoaromatics	0.236	0.301	0.369
2. Diaromatics	0.135	0.151	0.129
3. Triaromatics	0.094	0.086	0.055
4. Tetra-aromatics	0.081	0.055	0.074
5. Dibenzothiophene	0.020	0.020	0.020
6. Ether bridge	0.018	0.018	0.018
7. Biphenyl bridge	0.000	0.074	0.140
8. Alpha methyl	0.288	0.303	0.365
9. Beta methyl	0.082	0.060	0.098
10. Gamma methyl	0.096	0.096	0.096
11. Methylene bridge	0.004	0.088	0.067
12. Ethylene bridge	0.151	0.086	0.076
13. Hydroaromatic ring	0.117	0.162	0.140
14. Two ring hydroaromatic	0.049	0.046	0.040
15. Hydroaromatic bridge	0.064	0.039	0.046

[a]Functional group structures are shown in Figure VII-1.

Table VII-6
Functional Group Concentrations Derived from Mass Spectra

FUNCTIONAL GROUP CONCENTRATIONS DERIVED FROM MASS SPECTRA

Functional Group	High Resolution Data Concentrations in Moles/100 g	Low Resolution Data Relative Concentrations
1) (benzene ring)	.285	1
2) (naphthalene)	.150	.390
3) (phenanthrene/anthracene, 3 rings)	.096	.167
4) (pyrene, 4 rings)	.083	.174
5) (dibenzothiophene, O-S)	.010	—
6) (ether, —O—)	.034	—
7) (—•—)	.100	.323
8) —CH₃		—
9) ●—CH₃	.445	—
10) ○—CH₃		—
11) (—∧—)	.098	.135
12) (—∿—)	.082	.171
13) (cyclohexane ring)	.127	.356
14) (decalin, fused bicyclic)	.012	.029
15) (fused bicyclic)	.025	.038

Table VII-7

Comparison of Functional Group Analysis Results

with Average Parameter Calculations

	Average Parameter Calculations	Value Determined from Functional Group Concentrations
Aromatic C/Total C	0.72	0.70
Average number of rings per cluster	2.0	2.04
Nonbridge aromatic C/Total C	0.58	0.56
Naphthenic C/Total C	0.18	0.17
% Substitution of aromatic C	29.0	29.0

be examined. In the coal liquid application, functional group concentrations are estimated for a Solvent Refined Coal (SRC) utilizing elemental analysis, ^{1}H NMR and ^{13}C NMR data.

A heavy distillate obtained from a 50 ton/day SRC-II demonstration plant operating with Powhatan No. 5 coal was fractionated using a SARA chromatographic procedure (Jewell, 1974). The goal of the separation was to isolate heteroatomic functionalities. The ten primary fractions obtained were neutral oils, asphaltenes, very weak bases, weak bases, strong bases, very weak acids, weak acids, strong acids, neutral resins and saturates. Elemental composition, ^{1}H NMR and ^{13}C NMR data were obtained for the fractions. The data are given in Table VII-8.

For each fraction, the concentrations of the functional groups listed in Figure VII-2 were calculated. In some cases, functional groups were assigned concentrations of zero. For example, when separation procedures were specifically designed to exclude certain functionalities, e.g., phenolics in the neutral oil fraction of the SRC heavy distillate, the concentrations of those functionalities were set equal to zero. IR spectra were particularly helpful in determining the presence or absence of some functional groups in the samples.

Functional group concentrations were estimated for each of the ten SRC fractions; the results are given in Table VII-9. The concentrations were obtained by using the methodology described in Section VII.A. The elemental analysis and NMR data were used to construct the set of constraints.

$$\sum_{j=1}^{n} A_{ij}y_j = b_i \qquad (i=1,\ldots,m) \tag{15}$$

$$y_j \geq 0 \tag{16}$$

There were less constraints than unknown functional group concentrations (n>m) so the concentrations were determined by selecting the solution which had a distribution of aromatic ring sizes closest to that predicted by the method of Tominaga (1977). The distribution of ring sizes could have been included in the constraints of equation 15, however these data are more qualitative than the constraints of equation 15 and are therefore more appropriately used in the selection of a solution from the feasible space rather than in defining the feasible space. Thus, a solution was selected from the feasible space by minimizing equation 17 subject to the constraints of equations 15 and 16.

$$P = (C_i - 2y_2 - 4y_3 - 6y_4)^2 \tag{17}$$

where C_i is the concentration of internal carbon in the sample (Tominaga, 1977).

142

Table VII-8

Elemental analysis and NMR data for SRC-II heavy distillate fractions

Fraction	Yield (unnormalized wt% of heavy distillate)	Elemental Analysis Data (wt%)					^{1}H NMR data (wt%)				Carbon aromaticity
		C	H	N	O	S	Aromatic	Alpha	Beta	Gamma	
Neutral oils	73.4	91.4	7.2	0	0.72	0.66					
Aromatic fraction of neutral oils	69.3						49	28	17	6	0.81
Asphaltenes	8.4	85.4	6.4	3.2	4.5	0.47	59	28	10	3	0.81
Very weak bases	2.6	84.3	6.8	1.3	7.1	0.52	56	25	14	5	0.81
Weak bases	1.8	78.8	6.7	3.9	8.3	2.2	44	23	23	10	0.74
Strong bases	5.7	85.9	6.8	4.8	1.8	0.60	44	31	19	7	0.74
Very weak acids	1.0	82.2	8.0	0.21	8.9	0.16	48	32	16	4	0.74
Weak acids	1.2	82.3	7.7	0	9.8	0.16	58	25	11	4	0.79
Strong acids	0.2	76.0	5.6	0	17.8	0.60	53	32	11	3	0.79
Neutral resins	1.1	84.6	7.2	0.85	7.2	0.15	29	20	35	16	0.67
Saturates from neutral oils	4.2						0	0	74	26	0.00

Table VII-9

SRC-II Heavy Distillate Functional Group Concentrations (in moles/100 g)

functional[a] group	neutral oils	asphaltenes	very weak bases	weak bases	strong bases	very weak acids	strong acids	strong acids	neutral resins	whole heavy distillat.
1. dibenzothiophene	0.02	0.02	0.02	0.07	0.02	0.005	0.02	0.005	0.02	
2. carbazole	--	0.23	--	--	--	0.02	--	--	0.06	0.02
3. ether bridge	0.04	--	--	--	--	--	--	--	--	0.04
4. phenol	--	0.28	0.44	0.52	0.11	0.56	0.20	--	--	0.06
5. ketone	--	--	--	--	--	--	--	--	0.15	0.002
6. carboxylic acid	--	--	--	--	--	--	0.20	0.56	0.15	0.005
7. quinoline			--	0.28	0.17	--	--	--	--	0.02
8. aniline	--	--	0.09	--	0.17	--	--	--	--	0.01
9. benzene	0.03	0.41	0.25	0.32	0.30	0.76	0.79	0.14	0.03	0.11
10. naphthalene	0.34	0.006	0.10	0.003	0.09	0.002	0.02	0.04	0.001	0.27
11. hydroaromatic	0.13	0.001	0.001	0.04	0.001	0.055	0.001	0.001	0.001	0.10
12. α-methyl										
13. α-CH$_2$	0.07	0.005	0.34	0.48	0.22	0.56	0.67	0.60	0.68	0.11
14. α-CH										
15. β and ε (+) methyl	0.13	0.06	0.11	0.22	0.16	0.11	0.10	0.06	0.38	0.13
16. β and β (+) CH$_2$	0.05	0.001	0.001	0.27	0.001	0.16	0.20	0.12	0.80	0.06

Table VII-9 (continued)

functional[a] group	neutral oils	asphal tenes	very weak bases	weak bases	strong bases	very weak acids	strong acids	strong acids	neutral resins	whole heavy distillat.
17. β and β (+) CH										
18. phenanthrene	0.08	0.01	0.19	0.08	0.07	0.008	0.02	0.21	0.19	0.07
19. pyrene	0.04	0.005	0.02	0.009	0.05	0.006	0.01	0.02	0.05	0.04
20. methylene bridge	0.06	0.001	0.001	0.001	0.001	0.10	0.001	0.001	0.05	
21. ethylene bridge	0.18	0.14	0.05	0.001	0.05	0.08	0.001	0.001	0.001	0.16
22. biphenyl bridge	0.02	0.001	0.001	0.001	0.001	0.03	0.001	0.001	0.01	
23. two-ring hydroaromatic	0.04	0.007	0.10	0.08	0.09	0.05	0.03	0.03	0.04	0.04
24. two-ring hydroaromatic bridge	0.02	0.16	0.002	0.001	0.11	0.01	0.002	0.001	0.001	0.04
25. CH_2 in saturated molecules	0.41	--	--	--	--	--	--	--	--	0.31

a) Functional group structures are defined in Figure VII-2.

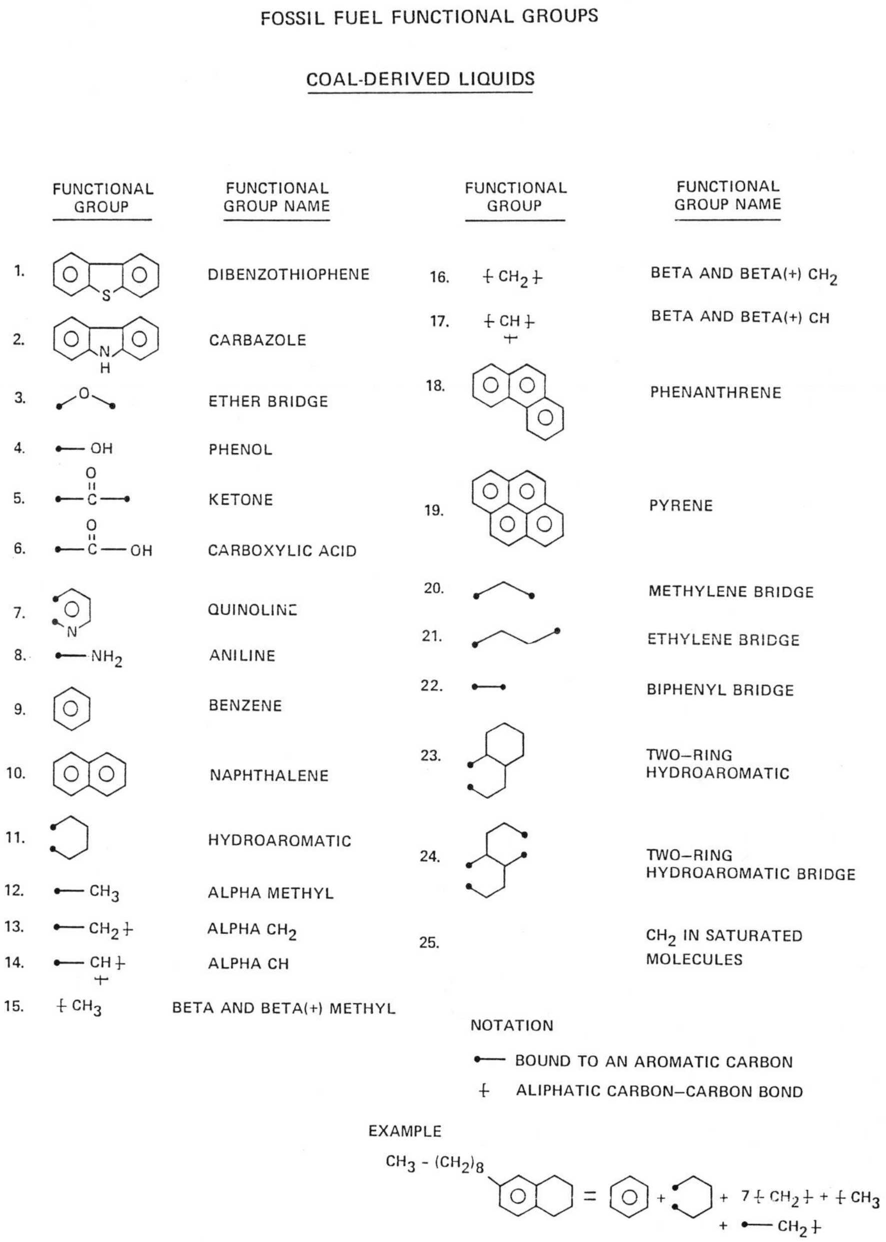

FOSSIL FUEL FUNCTIONAL GROUPS

COAL-DERIVED LIQUIDS

FUNCTIONAL GROUP	FUNCTIONAL GROUP NAME	FUNCTIONAL GROUP	FUNCTIONAL GROUP NAME
1.	DIBENZOTHIOPHENE	16. $+CH_2+$	BETA AND BETA(+) CH_2
2.	CARBAZOLE	17. $+CH+$	BETA AND BETA(+) CH
3.	ETHER BRIDGE	18.	PHENANTHRENE
4. $\bullet-OH$	PHENOL		
5.	KETONE	19.	PYRENE
6.	CARBOXYLIC ACID		
7.	QUINOLINE	20.	METHYLENE BRIDGE
8. $\bullet-NH_2$	ANILINE	21.	ETHYLENE BRIDGE
9.	BENZENE	22. $\bullet-\bullet$	BIPHENYL BRIDGE
10.	NAPHTHALENE	23.	TWO—RING HYDROAROMATIC
11.	HYDROAROMATIC	24.	TWO—RING HYDROAROMATIC BRIDGE
12. $\bullet-CH_3$	ALPHA METHYL		
13. $\bullet-CH_2+$	ALPHA CH_2	25.	CH_2 IN SATURATED MOLECULES
14. $\bullet-CH+$	ALPHA CH		
15. $+CH_3$	BETA AND BETA(+) METHYL		

NOTATION

$\bullet-$ BOUND TO AN AROMATIC CARBON

$+$ ALIPHATIC CARBON—CARBON BOND

EXAMPLE

$CH_3 - (CH_2)_8$

Figure VII-2. SRC Liquid Functional Groups

Examination of the functional group concentrations reveals that the neutral oils and the asphaltenes have a higher hydroaromatic content than the various acids and bases. High concentrations of aliphatic chains and small aromatic rings appear to be associated with high heteroatomic concentrations. Another interesting feature of the results is the ring size distribution. The neutral oils are predicted to contain mostly diaromatic rings, while the remainder of the fractions are predominantly monoaromatic.

The functional group concentrations of the whole heavy distillate were determined from the concentrations and yields of each of the fractions. The 25 concentrations, which are given in Table VII-9, provide a detailed structural profile of the heavy distillate and provide a reasonable starting point for the modeling the properties of this coal liquid.

In summary, this section has demonstrated the utility of functional group analysis in providing detailed profiles of complex, multi-component mixtures. The method can utilize a wide variety of data, and in general, is best suited for the structural analysis of complex liquids on which a large amount of data from a variety of sources is available. When applied, the method has several distinct advantages over other methods of structural characterization. It allows data from diverse sources such as NMR, IR and separation procedures to be incorporated into a single characterization; changes in structure are easily quantified; the types of structures present are easily visualized and the results could provide a starting point for property estimation.

VII.C APPLICATION TO HEAVY CRUDES

In this section we will apply Functional Group Analysis to the characterization of the atmospheric tower bottoms (ATBs) derived from a Maya crude oil. This example will differ from the preceding examples that dealt with coal liquids since the functional groups necessary to describe oils are slightly different than those required for coal liquids.

The atmospheric tower bottoms (ATB) of a Mayan crude oil were obtained by conventional distillation at 680°F. The elemental analysis data for the ATB are $C_7H_9O_{0.07}N_{0.04}S_{0.13}$. Proton and ^{13}C nuclear magnetic resonance spectra were obtained for the Maya ATB and were interpreted in the manner described in previous sections. These data are listed in Table VII-10.

Once again, to perform the characterization, the expected major functional groups present in the fuels must be defined. The functional groups proposed for the heavy oil were based on MS data, gas chromatography, and previous related results. The functional groups are shown in Figure VII-3.

Since the goal of the procedure is to estimate concentrations, the next step is to convert the available analytical data into a form which can be

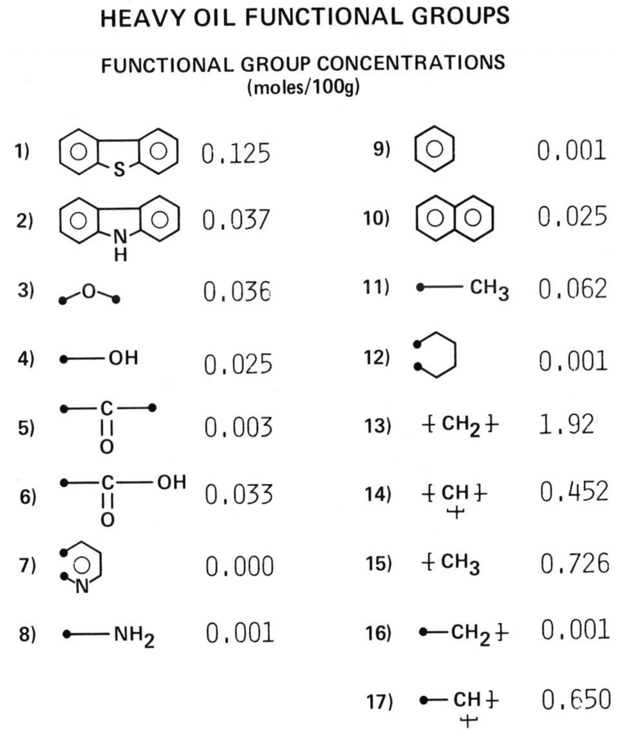

HEAVY OIL FUNCTIONAL GROUPS

FUNCTIONAL GROUP CONCENTRATIONS
(moles/100g)

1) [dibenzothiophene structure, S]   0.125

2) [carbazole structure, N-H]   0.037

3) —O—   0.036

4) •—OH   0.025

5) •—C(=O)—•   0.003

6) •—C(=O)—OH   0.033

7) [pyridine structure, N]   0.000

8) •—NH_2 0.001

9) [benzene ring]   0.001

10) [naphthalene structure]   0.025

11) •—CH_3 0.062

12) [cyclohexane structure]   0.001

13) $+CH_2+$ 1.92

14) $+CH+$ 0.452

15) $+CH_3$ 0.726

16) •—CH_2+ 0.001

17) •—$CH+$ 0.650

Figure VII-3.   Heavy Oil Functional Groups

Table VII-10

Elemental Analysis Data for Mayan ATB

%C	%H	%O	%N	%S
84.1	10.1	1.06	0.5	4.0

^{1}H NMR DAta for Mayan ATB

Hydrogen type	Chemical shift range (ppm from TMS)	% of Hydrogen
Aromatic hydrogen	9.0-5.0	7.7
Hydrogen in CH, CH_2 and CH_3 groups alpha to an aromatic ring	5.0-1.9	8.3
Hydrogen in CH and CH_2 groups beta or farther from an aromatic ring. Hydrogen in CH_3 groups beta to an aromatic ring	1.9-1.0	62.4
Hydrogen in CH_3 groups gamma or farther from an aromatic ring	1.0-5.0	21.6

^{13}C NMR Data for Mayan ATB

Carbon type	% of Carbon
Aromatic carbon	29.6
Carbon in CH groups	14.6
Carbon in naphthenic CH_2 groups Carbon in CH_2 groups alpha and gamma or farther from an aromatic ring	32.4
Carbon in CH_2 groups beta to an aromatic ring	5.2
Carbon in CH_2 groups next to a terminal methyl. Carbon in CH_2 groups beta to an aromatic ring in tetralin structures	3.5
Carbon in CH_3 groups alpha to an aromatic ring	2.7
Carbon in CH_3 groups attached to hydroaromatic structures	4.4
Carbon in CH_3 grups beta to an aromatic ring	2.4
Carbon in CH_3 groups gamma or farther from an aromatic ring	5.1

related to concentrations. For the sample considered in this work, the analytical data available included elemental analysis, ^{1}H NMR spectra, ^{13}C spectra, and separation yields.

The Maya ATB considered in this work is an overdetermined system since the amount of available data exceeds the number of functional groups proposed. Thus, a weighted least squares approach to estimating the functional group concentrations could be employed. The weighting factors would be related to the inverse of the uncertainty in the data. Uncertainties are not well known for NMR band assignments and mass spectra calibrations, however, so we weight the data by employing a modification of the method used for underdetermined systems, such as the coal liquids of section VII.B.

Functional group concentrations are estimated by using the elemental analyses and ^{1}H NMR data to construct a matrix of balance equations.

$$\sum_{i=1}^{n} A_{ij}y_j = b_i \qquad (i=1,\ldots,m) \tag{18}$$

$$y_j \geq 0 \tag{19}$$

These equations define a space of feasible solutions since n > m. A representative set of concentrations is selected from this space by choosing the solution which most closely matches the ^{13}C data. To do this, equation (20) is minimized over the space of feasible solutions.

$$\sum_{j=1}^{\ell} (x_j - f_j)^2 \tag{20}$$

where the x_j are the fractions of carbon in various bonding environments derived from the functional group concentrations and the f_j are the fractions of carbon derived from the ^{13}C spectrum. The functional group concentrations obtained in this manner for the Mayan ATB are given in Figure VII-3. These concentrations yield values of x_j which are in good agreement with the experimental f_j, as shown in Table VII-11.

The functional group concentrations shown in Figure VII-3 provide a concise and consistent representation of extensive analytical data.

For the Mayan ATB, the concentrations show that the degree of aromatic substitution is high and that the vast majority of the aromatic rings are heteroaromatic. More detailed interpretations and a comparison between heavy oils, tar sands, shale oils, and coal liquids are given in section VII.D.

VII.D STRUCTURAL PROFILES OF HEAVIER FRACTIONS OF FOSSIL FUELS

In this section, functional group analysis will be used to compare the structural profiles of heavy fractions of various fossil fuels. The fuels are

150

Table VII-11

Comparison of Carbon Distribution from Functional Group Concentrations and NMR

Carbon type	% of Carbon calculated from NMR spectrum	% of Carbon calculated from functional group concentrations
Aromatic carbon	29.6	26.9
Carbon in CH groups	14.6	14.3
Carbon in naphthenic CH_2 groups		
Carbon in CH_2 groups alpha and gamma or farther from an aromatic ring	32.4	32.6
Carbon in CH_2 groups beta to an aromatic ring	5.2	5.2
Carbon in CH_2 groups next to a terminal methyl; Carbon in CH_2 groups beta to an aromatic ring in tetralin structures	3.5	2.8
Carbon in CH_3 groups alpha to an aromatic ring	2.7	1.1
Carbon in CH_3 groups attached to hydroaromatic structures	4.4	5.6
Carbon in CH_3 groups to an aromatic ring	2.4	5.2
Carbon in CH_3 groups or farther from an aromatic ring	5.1	6.1

a Mexican crude atmospheric tower bottoms (ATB), a Canadian tar sand ATB, the heavy distillate of hydroliquefied bituminous coal, and the heavy distillate of a Colorado shale oil. Each of these fuels was separated by ion exchange chromatography into nine fractions. Analytical data from a variety of techniques, including elemental analysis, proton NMR, carbon-13 NMR and mass spectrometry, were integrated by the technique of functional group analysis. In addition to the expected differences due to heteroatom functionality, the technique of functional group analysis shows, quantitatively, the differences in the hydrocarbon skeletons between fractions of a given fuel and among the four fuels in a given fraction. Typically, these differences arise in the length of aliphatic chains, degree of branching, size of aromatic clusters, degree of aromatic substitution, and the occurrence of hydroaromatic groups.

VII.D.1 The Nature of the Fuels Being Compared

Analytical data on crude oils, tar sands, shale oil retorts and coal liquids were collected using ^{1}H NMR spectroscopy, ^{13}C NMR spectroscopy, mass spectrometry, elemental analysis and separation procedures. The fuels examined included an atmospheric tower bottoms from a Mayan crude oil, a Solvent Refined Coal-II heavy distillate, a Rio Blanco shale oil obtained with a modified in-situ retort and a crude derived from a Cold Lake Tar Sand.

The Maya, Cold Lake, SRCII heavy distillate and the Rio Blanco shale oil samples are all high boiling range materials. The Maya and Cold Lake samples used are the atmospheric tower bottoms (ATB), boiling above 360°C. The Rio Blanco shale oil is an ATB plus some 280-360°C range material, and the SRCII heavy distillate is the highest boiling, distillable portion of the product from the liquefaction of Powhatan no. 5 coal. The elemental data and boiling ranges for these samples are shown in Table VII-12.

In addition to the samples described above, referred to as whole samples, the Maya ATB, Cold Lake tar sand, Rio Blanco shale oil, and the SRCII heavy distillate were fractionated by extraction and ion exchange resin chromatography into asphaltenes, neutral oils, neutral resins, and acids and bases of various strengths using the technique developed by Jewell et al. (1974) and Ruberto et al. (1976), and applied to coal liquids by Petrakis et al. (1983b).

VII.D.2 Determination of Structural Profiles

The functional groups proposed for the heavy oils, tar sands, and coal derived liquids are based on MS data, gas chromatography, IR spectra and previous results. The functional groups are shown in Figure VII-4. The aliphatic functional groups proposed for the oils and tar sands are slightly different than those proposed for the SRCII heavy distillate. The reason for

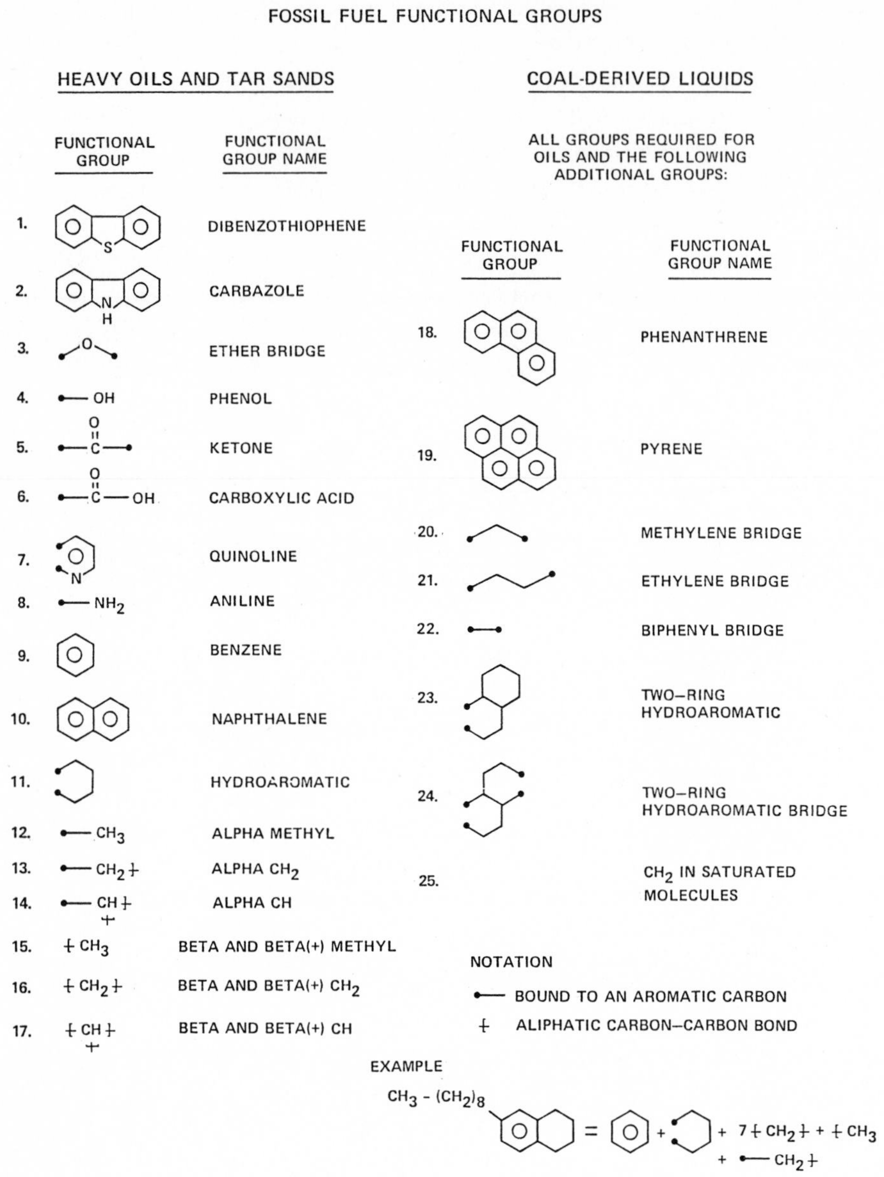

FOSSIL FUEL FUNCTIONAL GROUPS

HEAVY OILS AND TAR SANDS

	FUNCTIONAL GROUP	FUNCTIONAL GROUP NAME
1.		DIBENZOTHIOPHENE
2.		CARBAZOLE
3.	—O—	ETHER BRIDGE
4.	•—OH	PHENOL
5.	•—C— (=O) •	KETONE
6.	•—C—OH (=O)	CARBOXYLIC ACID
7.		QUINOLINE
8.	•—NH$_2$	ANILINE
9.		BENZENE
10.		NAPHTHALENE
11.		HYDROAROMATIC
12.	•—CH$_3$	ALPHA METHYL
13.	•—CH$_2$+	ALPHA CH$_2$
14.	•—CH+ +	ALPHA CH
15.	+CH$_3$	BETA AND BETA(+) METHYL
16.	+CH$_2$+	BETA AND BETA(+) CH$_2$
17.	+CH+ +	BETA AND BETA(+) CH

COAL-DERIVED LIQUIDS

ALL GROUPS REQUIRED FOR OILS AND THE FOLLOWING ADDITIONAL GROUPS:

	FUNCTIONAL GROUP	FUNCTIONAL GROUP NAME
18.		PHENANTHRENE
19.		PYRENE
20.		METHYLENE BRIDGE
21.		ETHYLENE BRIDGE
22.	•—•	BIPHENYL BRIDGE
23.		TWO–RING HYDROAROMATIC
24.		TWO–RING HYDROAROMATIC BRIDGE
25.		CH$_2$ IN SATURATED MOLECULES

NOTATION

•— BOUND TO AN AROMATIC CARBON

+ ALIPHATIC CARBON–CARBON BOND

EXAMPLE

$$CH_3 - (CH_2)_8$$

= ⬡ + ⬡ + 7 +CH$_2$+ + +CH$_3$ + •—CH$_2$+

Figure VII-4.   Functional Groups for Heavy Fuels

Table VII-12

Analyses of Heavy Fossil Fuels

| Material | Boiling Range (°C) | Weight Percent | | | | | |
		C	H	O	N	C	Total
Cold Lake Tar Sand	360+	84.0	10.0	0.2	0.4	5.1	99.7
Maya Heavy ATB	360+	84.3	10.4	0.5	0.5	4.7	100.4
SRCII Heavy Distillate	288-482	89.5	7.7	2.3	1.1	0.4	101.0
Rio Blanco	288+	83.6	11.3	0.6	1.8	1.8	99.1

this difference is that only a small fraction of the aliphatic structures present in the SRCII heavy distillate are aliphatic chains, while the majority of the aliphatic carbon present is part of hydroaromatic groups or aliphatic bridges. In heavy oils, tar sands and shale oils, aliphatic chains dominate over hydroaromatic groups and bridges. For these oils, a detailed structural characterization need not include bridges and large hydroaromatics that are necessary for coal liquids. Thus, the major aliphatic groups proposed for the oils differ from those proposed for the coal liquid.

The Solvent Refined Coal and Cold Lake Tar sands samples are underdetermined systems and are treated in the manner described in section A of this chapter. The oil samples considered in this work are overdetermined systems since the amount of available data exceeds the number of functional groups proposed. Not all of the sources of data are equally accurate, however, so in the optimal case a weighted least squares procedure would be employed with the weights set equal to the inverse of the uncertainty in the data. Uncertainties are not well known for NMR band assignments and mass spectra calibrations, so we weight the data by employing a modification of the method used for underdetermined systems. The elemental analysis, ^{1}H NMR and separation data are used to define the matrix of equations. The ^{13}C and mass spectral data are then used in a least squares minimization using one of the three methodologies defined below. Different methods are used only because the types of data available on each of the samples are different.

Method one requires separation yields, elemental analysis, ^{1}H NMR and ^{13}C NMR data for each fraction. Functional group concentrations are estimated for each fraction by using the elemental analyses and ^{1}H NMR data to construct a matrix of balance equations. These equations define a space of feasible solutions. A representative set of concentrations is selected from this space by choosing the solution which most closely matches the ^{13}C data. To do this, equation (21) is minimized over the space of feasible solutions.

$$\sum_{j=1}^{\ell} (x_j - f_j)^2 \tag{21}$$

where the x_j are the fractions of carbon in various bonding environments derived from the functional group concentrations and the f_j are the fractions of carbon derived from the ^{13}C spectrum. The ^{13}C spectra are used in this qualitative manner due to the uncertainties associated with overlapping band assignments.

The second method for estimating concentrations requires separation yields and elemental analysis data for each of the fractions and ^{1}H NMR and ^{13}C NMR data on the whole oil. A feasible space of solutions is defined by the heteroatomic concentrations, the ^{1}H NMR data on the whole fuels and the

elemental analysis data on the whole fuels. A representative set of concentrations is chosen by selecting the solution which most closely matches the
^{13}C data on the whole fuel.

The third method for estimating concentrations utilizes separation yields,
mass spectra, elemental analysis and ^{1}H NMR data. Estimates of functional
group concentrations derived from mass spectra are used to select a set of
concentrations from the feasible space defined by the separation yields, ^{1}H
NMR and elemental analysis data. The method is identical to method one,
except that equation 21 is replaced by equation 22.

$$\sum_{j=1}^{n} (y_j - h_j)^2 \qquad\qquad (22)$$

where y_j is the concentration of functional group j and h_j is the concentration of functional group j determined from the mass spectra. The details of
this method are described in section VII.A.

Method one was applied to each of the Rio Blanco fractions and the results
are given in Table VII-13. The functional group concentrations for the SRCII
fractions are reported in Table VII-9. Separation yields were then used to
estimate functional group concentrations in the whole fuels. These concentrations are listed in Table VII-14. Methods two and three were applied to the
Mayan whole oil. The results are reported in Table VII-14. The major structural features of the fuels are compared and discussed below.

VII.D.3 Comparison of Structural Profiles

We first compare the separation yields for the various fuels. The separation yields from the Maya and Cold Lake ATBs, the oil shale, and the coal
liquid heavy distillate are shown in Figure VII-5. The neutral oil yields are
similar for the four materials, but the asphaltene yields show large variation. The Rio Blanco shale oil has the lowest concentration of asphaltenes,
followed by the coal liquid. The crude and tar sand yield more than three
times the weight of asphaltenes as the coal liquid distillate. The coal
liquid and the oil shale have a much higher concentration of bases with most
of the basic material in the oil shale being in the weak base fraction and
most of the basic material in the coal liquid being strong bases. Shale oil
has the most neutral resins; the coal liquid distillate, the least; and the
crude and the tar sand were intermediate. In the acid yields, Rio Blanco
shale oil has the most strong acids, SRCII heavy distillate, the least. The
Maya crude and the Cold Lake tar sands have very few weak acids compared to
the shale oil and the coal liquid distillate. In general, the Rio Blanco
shale oil had the most O and N containing material, followed by the coal
liquid, with the Maya crude having the least acid, base, or neutral resin

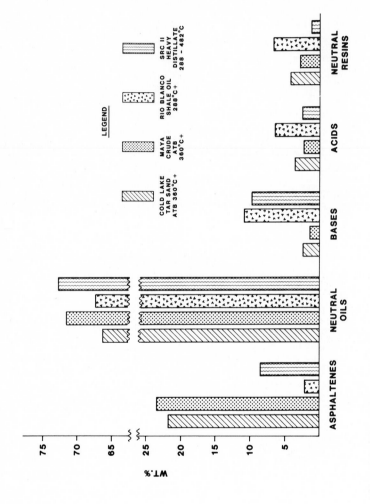

Figure VII-5. Separation Yields of Heavy Fossil Fuels

Table VII-13

Rio Blanco Shale Oil Functional Group Concentrations (in mol/100 g)

functional group[a]	neutral oils	very weak bases	weak bases	strong bases	very weak acids[b]	neutral resins
1. dibenzothiophene	0.04	0.05	0.11	0.11	NA	0.04
2. carbazole	<0.001	--	--	--	NA	0.10
3. ether bridge	0.06	--	--	--	--	--
4. phenol	--	0.35	0.48	1.25[c]	NA	--
5. ketone	--	--	--	--	--	0.16
6. carboxylic acid	--	--	--	--	--	0.16
7. quinoline	--	--	0.19	0.09	NA	--
8. aniline	--	0.19	--	0.09	--	--
9. benzene	0.002	0.01	0.10	<0.001	0.13	0.05
10. naphthalene	0.07	0.23	0.002	<0.001	0.19	0.001
11. hydroaromatic	0.002	0.01	0.003	0.01	0.002	0.009
12. α-methyl	0.07	0.08	0.58	0.03	0.02	0.37
13. α-CH$_3$	0.35	0.38	0.32	0.06	0.24	0.07
14. α-CH	0.20	0.55	0.03	0.13	0.94	0.01
15. β and β (+) methyl	1.16	0.66	0.48	0.74	0.77	0.88
16. β and β (+) CH$_2$	3.99	2.08	2.05	2.63	2.10	2.99
17. β and β (+) CH	0.15	0.07	0.47	0.01	0.04	0.13

[a]Functional group structures are shown in Figure VII-4. [b] 13C data on the whole oil were used in estimating these concentrations. [c]These heteroatomic groups are undoubtedly bound to both aromatic and aliphatic carbons. NA = Not Applicable.

Table VII-14

Functional Group Concentrations of Whole Fossil Fuels (Concentrations in mol/100 g)

functional group[a]	SRC heavy distillate	Rio Blanco shale oil	Cold Lake tar sands	Mayan atmospheric tower bottoms	
				method 2[b] calcns	method 3[b] calcns
1. dibenzothiophene	0.02	0.05	0.15	0.12	0.12
2. carbazole	0.02	>0.01	0.03	0.03	0.03
3. ether bridge	0.04	0.04	0.03	0.04	0.04
4. phenol	0.06	0.09	>0.09	0.03	0.03
5. ketone	0.002	0.01	0.001	0.003	0.003
6. carboxylic acid	0.005	0.02	>0.02	0.03	0.03
7. quinoline	0.02	>0.02	<0.001	<0.001	<0.001
8. aniline	0.01	>0.01	0.002	0.001	0.001
9. benzene	0.15	0.01	0.002	0.05	0.02
10. naphthalene	0.27	0.06	<0.001	0.01	0.03
11. hydroaromatic	0.10	0.003	0.07	0.01	0.02
12. α-methyl		0.13	0.009	0.01	0.002
13. α-CH$_2$	0.11	0.31	0.23	0.07	0.01
14. α-CH		0.18	0.24	0.62	0.69
15. β and β (+) methyl	0.13	1.06	1.01	0.73	0.73

Table VII-14 (continued)

functional group[a]	SRC heavy distillate	Rio Blanco shale oil	Cold Lake tar sands	Mayan atmospheric tower bottoms	
				method 2[b] calcns	method 3[b] calcns
16. β and β (+) CH₂	0.06	3.67	2.03	2.94	3.04
17. β and β (+) CH		0.18	0.97	0.40	0.10
18. phenanthrene	0.07	--	--	--	--
19. pyrene	0.04				
20. methylene bridge	0.05	--	--	--	--
21. ethylene bridge	0.16	--	--	--	--
22. biphenyl bridge	0.01	--	--	--	--
23. two-ring hydroaromatic	0.04	--	--	--	--
24. two-ring hydroaromatic bridge	0.04	--	--	--	--
25. CH₂ in saturated molecules	0.31	--	--	--	--

a Functional group structures are shown in Figure VII-4.
b Methods 2 and 3 are defined in the text.

material. The significance of these differences in yield, in terms of concentration of chemical structures, are made more clear by the results of the functional group determination.

The functional group concentrations of Table VII-14 reveal that the differences in separation yields among the various fuels result in significant differences in the concentration of phenols, carboxylic acids, amines, quinolines and dibenzothiophenes. Ether and carbozole concentrations remain relatively constant in all of the fuels.

A striking difference between the structural characterization of coal liquids and the structural profiles of the oils and tar sands is the concentration of non-heteroatomic rings. Virtually all of the aromatic ring systems in the oils and tar sands appear to be heteroatomic while the majority of the rings in coal liquids do not contain heteroatoms. This result implies that upgrading processes for coal liquids must be much more selective than processes for tar sands and oils. If coal liquid processes are not highly selective, then hydrogen consumption will be very high, possibly as much as 30 atoms of hydrogen per heteroatom removed (Petrakis, 1983a).

Another feature worthy of note in the structural profiles is the difference in hydrocarbon structure of the various fractions. It seems that the fractions do not merely differ by the presence or absence of certain heteroatomic groups. Additional differences appear in the hydrocarbon skeletons. For example, in the SRCII heavy distillate the asphaltenes, very weak acids and weak acids have much smaller average aromatic ring sizes than the other fractions. The acids have higher concentrations of aliphatic chains and low concentrations of aliphatic bridges. As another example, consider the Rio Blanco retort fractions. The very weak bases appear to have much longer chains than the other fractions (compare the number of CH_2 carbons to the number of terminal methyls and points of attachment to aromatic rings).

The significance of these variations in hydrocarbon structure between the various fractions is the implied conclusion that certain heteroatomic groups are associated with certain hydrocarbon groups. These associations are not precise or well defined yet, but they might be exploited in developing selective upgrading processes.

The uncertainty associated with the functional group concentrations can be roughly estimated by comparing the results of the two different methods for estimating Mayan ATB concentrations. As shown in Table VII-14, the concentrations based on elemental analyses, separation yields, ^{1}H NMR and ^{13}C NMR data were in reasonable agreement with concentrations obtained using elemental analyses, separation yields, ^{1}H NMR and mass spectra. Previous results (Section VII.A and Petrakis, 1983a) have shown that functional group concentrations estimated from elemental analysis, ^{1}H NMR and mass spectral data were in good agreement with the actual concentrations present in model compound mixtures.

Thus, structural profiles can be generated for a variety of fossil fuels by estimating the concentrations of the major functional groups present in the fuels. This type of structural characterization allows the systematic incorporation of a large number of data into a concise form and serves as a starting point for making structural comparisons and as a basis for estimating thermodynamic and physical properties.

REFERENCES

Allen, D. T., Gray, M. R., and Le, T. T., Liquid Fuels Technology, $\underline{2}$, 327 (1984).

Allen, D. T., Grandy, D. W., Jeong, K. M. and Petrakis, L., Ind. Eng. Chem. Process Des. Dev., $\underline{24}$, 737 (1985).

Clutter, D. R., Petrakis, L., Stenger, R. L. and Jensen, R. K., Anal. Chem., $\underline{44}$, 1395 (1972).

Herod, A. A., Ladner, W. R., and Shape, C. E., Philos. Trans. R. Soc. London, Ser. A, $\underline{300}$, 3 (1981).

Jewell, D. M., Albaugh, E. W., Davis, B. E. and Ruberto, R. G., Ind. Eng. Chem. Fundam., $\underline{13}$, 178 (1974).

Jewell, D. M., Albaugh, E. W., Davis, B. E., and Ruberto, R. G., Ind. Eng. Chem. Fundam., $\underline{22}$, 292 (1983).

Le, T. T. and Allen, D. T., Fuel, $\underline{64}$, 1754-1759 (1985).

Luus, R., and Jaakola, T. H. I., AIChE J., $\underline{19}$, 760 (1973).

Lydersen, A. L., Univ. Wisconsin Coll. Eng. Eng. Exp. Sta., Rep. 3, 1955.

Petrakis, L., Allen, D. T., Gavalas, G. R., Gates B. C., Anal. Chem., $\underline{55}$, 1557 (1983a).

Petrakis, L., Ruberto, R. G., Young, D. C., Gates, B. C., Ind. Eng. Chem. Process Des. Dev., $\underline{22}$, 292 (1983b).

Ruberto, R. G., Jewell, D. M., Jensen, R. K. and Cronauer, D. C., Adv. Chem. Ser., No. 151, Chapter 3, 1976.

Tominaga, H., Seiji, I., and Yashiro, M., Bull. Jpn. Pet. Inst., $\underline{19}$, 50 (1977).

Chapter VIII

ESTIMATION OF PROPERTIES FROM NMR CHARACTERIZATIONS

To this point, we have considered NMR techniques and methods for using NMR data to characterize the structures of liquid fuels. We have seen that the structural characterizations are useful in their own right. They provide a basis for quantitatively comparing the structural features of very different fuels. In this section we will take the utilization of our structural characterization one step farther. We will demonstrate that the structural analyses, and in particular the functional group characterization, can be used to reliably predict the thermophysical properties of liquid fuels.

Structural characterizations can be incorporated into property estimation schemes in a variety of ways. Group additivity principles and the concept of continuous thermodynamics are both useful tools in this regard. Group additivity methods make the assumption that each chemical functional group contributes a definite value to the property of the mixture, regardless of how the groups associate themselves into molecules (Luria, 1977; Fredenslund, 1975). Thus, in the simplest case, the property of a mixture can be estimated by multiplying the concentration of each group by the group's contribution. So, this method of estimating properties is easily interfaced with any characterization method that specifies group concentrations. The methods of continuous thermodynamics (Cotterman, 1985) require that a characterization function $F(\lambda)$ be available. The form of $F(\lambda)$ is flexible, as is the choice of independent variable λ, so this approach can be used in conjunction with a variety of characterization methods. A third method for incorporating characterization data into property estimation is to develop property correlations using characterization variables as independent parameters.

Consider now how these property estimation methods will interface with the characterization methods discussed in the previous chapters.

Average Molecule Characterizations

Characterizing a mixture by constructing an average molecule is equivalent to assuming that the properties of the mixture can be approximated by the properties of the average molecule. If the average molecule is a stable species, property data may exist for it. Otherwise, the properties of the average molecule, and thus the mixture, can be estimated using group additivity methods. An example of this approach is provided by Ruzicka (1983) who characterized petroleum fractions by assigning an average structure to each fraction, then applied group contributions methods (UNIFAC, Fredenslund, 1975) to estimate the vapor pressure.

Average Structural Parameters

Average structural parameters can be interfaced with property estimation methods either through correlations or through the methods of continuous thermodynamics. For example, structural parameters based on elemental composition and NMR data have been used as variables in correlations for equation of state parameters (Alexander, 1985a,b).

Functional Group Characterization

Functional group characterizations have been interfaced with property estimation methods in two ways. First, group concentrations can be used in conjunction with group contribution methods to estimate bulk thermodynamic properties (Le and Allen, 1985). An example of this approach is described in detail below.

The second method of utilizing this type of characterization is simply a variation of the method used for average molecules. Given a functional group distribution and an average molecular weight, drawing an average molecule is a straightforward calculation. This average molecule will have non-integer group concentrations, but the properties of this average molecule can still be calculated using group contribution methods. For most group contribution methods non-integer concentrations do not present a problem. In addition, allowing non-integer group concentrations makes the construction of average molecules a more stable calculation. Small perturbations in the data are much less likely to produce dramatic changes in the average structure when non-integer values are allowed.

As a case study of interfacing structural characterizations with property estimation methods, we consider the heat capacity of coal derived liquids. Specifically, we will predict the liquid heat capacity, as a function of temperature, for a Solvent Refined Coal (SRC) Heavy Distillate using a functional group approach. It should be noted, however, that heat capacity is only one of the many thermodynamic properties that can be estimated using this type of approach.

VIII.A THERMODYNAMIC PROPERTIES

VIII.A.1 Heat Capacities

To underscore the significance of combining structural characterizations with property estimation methods we will briefly review the methods used to estimate thermodynamic properties in one type of complex mixture, i.e., coal liquids. Since the property we have chosen to demonstrate this approach is heat capacity, this brief review will focus on that property. A comprehensive review of the correlations available for predicting liquid heat capacities and other thermodynamic properties has been given by Reid (1977).

Correlations designed to predict the heat capacities of coal liquids generally fall into one of three broad categories.

1.   corresponding states methods;

2.   curve fitting methods;

3.   group contribution methods.

The principle underlying corresponding state correlations is that at the same reduced temperature, the liquid fuel will behave in the same way as a reference fluid. The problems with this type of approach for complex liquid fuels are twofold. First, defining a critical temperature for a multicomponent liquid fuel fraction is difficult. Second, a single reference fluid is not generally adequate to mimic the intermolecular interactions occurring in heavy fuels. A corresponding states approach does allow for a very consistent method of estimating thermodynamic properties of liquid fuels, however. Thus it remains a popular approach. Starling (1981) proposed a multiparameter equation of state based on corresponding state theory for predicting the thermodynamic and physical properties of coal liquids. The expression for the compressibility factor, z, is given by:

$$z = z_0(T^*, \rho^*) + \Upsilon z_\Upsilon(T^*, \rho^*) \tag{1}$$

where z_0 and z_Υ are universal functions of the reduced density, ρ^*, and the reduced temperature, T^*, and Υ is the orientation parameter. Complete expressions for the equation of state and empirical correlations for these parameters are given in Starling (1981). Thermodynamic properties can then be derived from the equation of state by classical thermodynamic relations (Reid, 1977).

Curve fits are also frequently used to estimate the properties of liquid fuels. A typical correlation is that proposed by Watson and Nelson for the heat capacity of petroleum derived fuels, shown in equation 2.

$$C_p = (0.35 + 0.055 K_w) \times [0.6811 - 0.308S + (0.815 - 0.306S)\frac{t}{1000}] \tag{2}$$

These correlations require the specific gravity, S, and a characterization parameter, K_w, to predict heat capacity as a function of temperature, t. A modified form of the Watson-Nelson equation for use with coal liquids was developed by Mraw (1984) and is given below.

$$C_p = (0.06759+0.05638K_w) \times [0.6450-0.05959S+(1.2892-0.5264S)\frac{T}{1000}] \qquad (3)$$

where K_w is the Watson Characterization Factor ($K_w = [T_b(^\circ R)]^{1/3}/S$), T_b is the boiling point, S is the $60^\circ/60^\circ$ specific gravity and T is the temperature, in $^\circ F$, at which the heat capacity is to be evaluated.

Group contribution methods are based on the premise that each group contri‑ butes a definite value to the heat capacity of the mixture, regardless of how the groups associate themselves into molecules. Thus, to estimate a mixture's heat capacity using group contribution methods, it is only necessary to know the concentrations of the functional groups present. Group contribution approaches for estimating heat capacities of coal liquids and other liquid fuels have been neglected until recently because of the lack of a reasonable method for estimating group concentrations. As described in previous chapters, a methodology is now available for estimating group concentrations in complex mixtures, and with this tool, highly accurate and flexible correla- tions can be formulated for liquid heat capacities of polar, multicomponent mixtures such as coal liquids.

The three basic methods for estimating liquid heat capacity have been applied to a set of heavy fuels. These fuels are the heavy distillate frac‑ tion and narrow boiling fractions of the Solvent Refined Coal described in Chapter VII and by Allen (1984). Separation yields, ^{1}H NMR spectra, ^{13}C NMR spectra, mass spectra and elemental analysis data were available for the heavy distillate. Some of these data are reported in Table VIII-1. A more complete discussion of sample generation procedures, analytical characterization methods and additional analytical data are given by Gray (1981) and Gray and Holder (1982) and Allen (1984,1985).

The application of the corresponding states method and Mraw's method for estimating heat capacity is discussed in detail elsewhere (Gray and Holder, 1982). For the group contribution method, the first step was to estimate functional group contributions. Functional group concentrations in the coal liquids were estimated using the methods described in previous chapters. The procedure is essentially a least squares fit of the functional group concen- trations to the analytical data. A representative set of functional groups was chosen for the coal liquid fractions based on the available analytical data, and the groups are shown in Figure VIII-1. For the narrow boiling range fractions the functional group concentrations were forced to exactly satisfy the elemental analysis data. A set of 5 linear equations was constructed to represent balances on the total concentrations of carbon, hydrogen, oxygen, nitrogen and sulfur. Since each element is distributed among the several

167

Table VIII-1

Characteristics of SRC-II Coal Liquid Fractions

Fraction	Boiling Range (°F)	Elemental Analysis (Wt %)				
		C	H	N	O	S
No. 6	374-399	83.6	9.5	0.82	6.1	0.07
No. 10	538-598	88.3	9.0	0.76	1.8	0.17
Heavy Distillate	550-900	85.7	6.8	0.66	1.7	0.62

COAL LIQUID FUNCTIONAL GROUPS

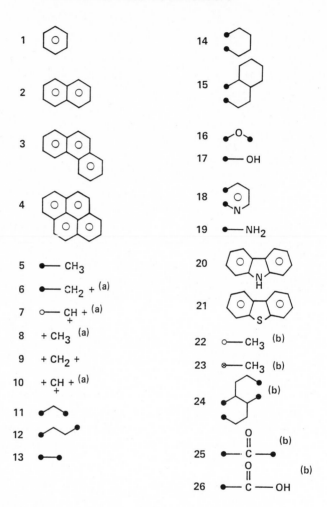

NOTATION:

●— BOUND DIRECTLY TO AN AROMATIC RING

○— BOUND TO A CARBON ALPHA TO AN AROMATIC RING

⊗— BOUND TO A CARBON BETA OR FURTHER FROM AN AROMATIC RING

+   ALIPHATIC CARBON – CARBON BOND

Figure VIII-1.   Coal Liquid Functional Groups. (a) Group assigned to narrow boiling functions only; (b) group assigned to heavy distillate only. (Note that group 16 is probably a thermal degradation product of the Dowtherm used in the Fort Lewis SRC-II plant).

functional groups, these equations are simple elemental balances of the form,

$$\sum_{j=1}^{n} A_{ij}y_j = b_i \qquad (j=1,\ldots m) \qquad (4)$$

where y_j represents the concentration of functional group j; the b_i are the concentrations of the various atomic types, as determined by elemental analysis, and A_{ij} are the stoichiometric coefficients relating group j to atomic type i. To be physically meaningful, the concentrations must be non-negative,

$$y_j \geq 0 \qquad (5)$$

A similar procedure was used to estimate the concentrations of the heavy distillate. More group concentrations are reported for this sample because more extensive analytical data were available. Equations similar to 4 and 5 were developed using NMR, elemental analysis and separation data. The full details of the estimation are given in Chapter VII. The results of the calculations are reported in Table VIII-2.

For the group contribution method, the contribution of each group to the heat capacity of the mixture is expressed as a third order polynomial

$$C_{p(\ell)} = A+BT+CT^2+DT^3 \qquad (6)$$

The constants A, B, C and D for each coal liquid functional group are given in Table VIII-3. The values for hydrocarbons groups were based on the work of Luria and Benson (1977) and the values for heteroatomic groups were based on the results of Missenard (1965) and Reid (1976). The sources for hydrocarbon and heteroatomic groups sources differ slightly in that Missenard's results do not incorporate the effects of nearest neighbors, while Benson and Luria's results do. Ignoring the effects of nearest neighbors results in slightly less accuracy, but since the concentrations of heteroatoms are significantly less than the hydrocarbon concentrations, the use of Missenard's data is a useful first approximation. The values of A, B, C, and D for the coal liquids were determined by multiplying the group value for each of the constants by the group concentration, then summing over all groups.

Heat capacity predictions were compared directly with experimental data for the narrow boiling fractions. For the heavy distillate fraction, no heat capacity data were available. Heat capacity values were estimated by multiplying the fraction of the heavy distillate boiling off in a narrow temperature range by the heat capacity of that narrow boiling fraction (Gray, 1981; Gray and Holder, 1982), then summing over the entire distillation curve for the heavy distillate. The results are plotted in Figures VIII-2 to VIII-4.

Table VIII-2

Functional Group Concentrations

Functional Group[a]	Concentrations in Moles/100 g		
	Fraction #6	Fraction #10	Heavy Distillate
1. Benzene	.630	.327	.154
2. Naphthalene	0	.202	.273
3. Phenanthrene	0	0	.073
4. Pyrene	0	0	.036
5. Alpha Methyl	.055	.030	.108
6. Alpha CH_2	0	.066	-
7. Alpha CH	0	0	-
8. Aliphatic Methyl	.302	.133	-
9. Aliphatic CH_2	.836	.334	.309
10. Aliphatic CH	.198	.072	-
11. Methylene Bridge	0	.020	.047
12. Ethylene Bridge	0	.008	.155
13. Biphenyl Bridge	0	.007	.014
14. Hydroaromatic	.233	.284	.104
15. Two-ring Hydroaromatic	.108	.188	.041
16. Ether Bridge	0	.004	.035
17. Phenol	.382	.105	.062
18. Quinoline	0	0	.016
19. Aniline	.061	.050	.013
20. Carbazole	0	0	.021
21. Dibenzothiophene	.002	.009	.019
22. Beta Methyl	-	-	.058
23. Gamma(+) Methyl	-	-	.128
24. Two-ring Hydroaromatic Bridge	-	-	.039
25. Ketone	-	-	.002
26. Carboxylic Acid	-	-	.005

(a) Functional group structures are shown in Figure VIII-1.

Table VIII-3

Group Contributions to Liquid Heat Capacity at Constant Pressure

Functional Group[a]	$C_{p(l)}$ (group) = A + BT + CT^2 + DT^3 Group Contribution in cal/mol K			
	A	B	C	D
1. Benzene	-1.105E1	3.467E-1	-1.030E-3	1.197E-6
2. Naphthalene	-2.230E1	5.135E-1	-1.349E-3	1.401E-6
3. Phenanthrene	-3.354E1	6.803E-1	-1.668E-3	1.604E-6
4. Pyrene	-4.110E1	7.316E-1	-1.645E-3	1.409E-6
5. Alpha Methyl	3.911E1	-3.381E-1	1.093E-3	-1.130E-6
6. Alpha CH$_2$	6.084E1	-6.214E-1	2.152E-3	-2.418E-6
7. Alpha CH	3.314E1	-3.864E-1	1.468E-3	-1.759E-6
8. Aliphatic Methyl	8.459E0	2.113E-3	-5.605E-3	1.723E-7
9. Aliphatic CH$_2$	-1.383E0	7.049E-2	-2.063E-4	2.269E-7
10. Aliphatic CH	2.489E0	-4.617E-2	3.181E-4	-4.565E-7
11. Methylene Bridge	7.103E1	-7.904E-1	2.851E-3	-3.290E-6
12. Ethylene Bridge	1.217E2	-1.243E0	4.303E-3	-4.835E-6
13. Biphenyl Bridge	-3.876E0	-6.430E-2	3.670E-4	-5.945E-7
14. Hydroaromatic	1.189E2	-1.102E0	3.890E-3	-4.381E-6
15. Two-ring Hydroaromatic	8.483E1	-6.730E-1	2.869E-3	-3.462E-6
16. Ether Bridge	5.910E0	0.400E-2	-	-
17. Phenol	-7.550E0	3.660E-2	0.790E-4	-
18. Quinoline	4.736E0	1.412E-1	-3.313E-4	3.013E-7

Table VIII-3 (continued)

Functional Group (a)	Group Contribution in cal/mol K $C_{p(\ell)}$ (group) = $A + BT + CT^2 + DT^3$			
	A	B	C	D
19. Aniline	3.905E1	-2.004E-1	4.000E-4	—
20. Carbazole	-1.378E1	6.291E-1	-1.692E-3	1.800E-6
21. Dibenzothiophene	-1.880E1	6.359E-1	-1.692E-3	1.800E-6
22. Beta Methyl	3.019E1	-2.812E-1	1.002E-3	-1.115E-6
23. Gamma(+) Methyl	-1.383E0	7.049E-2	-2.063E-4	2.269E-7
24. Two-ring Hydroaromatic Bridge	1.758E2	-1.818E0	6.752E-3	-7.826E-6
25. Ketone	6.931E1	-6.724E-1	2.299E-3	-2.605E-6
26. Carboxylic Acid	3.600E1	-2.944E-1	1.150E-3	-1.303E-6

(a) Functional group structures are shown in Figure VIII-1.

Since the narrow boiling fractions and the heavy distillate were obtained using the same coal and the same process, the heat capacity estimates should be reliable.

The results, shown graphically in Figures VIII-2 to VIII-4, and listed in Table VIII-4 are in excellent agreement with the data, despite a slight negative bias. Table VIII-5 shows that the group contribution results are more accurate than both the Starling and the modified Watson-Nelson corresponding states correlations over the range of applicability of the group contribution expressions. The results are also more accurate than values read from the graphical correlation presented in the SRC-II Process Physical Properties Data Book. The slight negative bias in the results is most pronounced in the samples with high phenolic concentrations. This suggests that the bias may be due to insufficient corrections for hydrogen bonding. The hydrogen bonding that is not accounted for by the group contribution method is probably due to interactions between different functional groups. The group contribution approach is unable to account for these interactions since the group contributions are based on pure compound data which generally only include self associations. Deviations in the results for the heavy distillate fraction might be due to errors in estimating the observed heat capacity, since the boiling ranges of some narrow boiling fractions making up the distillate overlapped.

The accuracy of the group contribution method begins to break down significantly when temperatures exceed 400 K, the maximum temperature used in determining the group contribution values, A, B, C and D. Beyond the range of applicability of the group contribution method, heat capacities can be estimated either by linearly extrapolating the liquid phase group contribution equations or by using the ideal gas group contribution values given by Benson (1976). The ideal gas values result in a negative bias which reaches a maximum magnitude of roughly 20%, as shown in Table VIII-6. The linear extrapolations are much more accurate; typically producing errors of less than 5%. The linear coefficients (C_p = A+BT) are given in Table VIII-7.

It is instructive at this point to determine whether a detailed chemical characterization is really necessary to estimate the heat capacity. By comparing the results using Mraw's correlation (equation 2) with the group contribution approach it is apparent that, at least for the case of coal liquids, detailed structural data are quite helpful in providing accurate heat capacity estimates for samples in which the concentration of hydrogen bonding species is high. An additional issue that needs to be addressed is which sources of analytical data are of the most value in estimating these heat capacities. Recent work suggests that elemetal analysis and ^{1}H NMR data are sufficient for accurate estimates of coal liquid heat capacities. The elemetal analysis data are used in the characterization to estimate the concentrations of heteroatoms

174

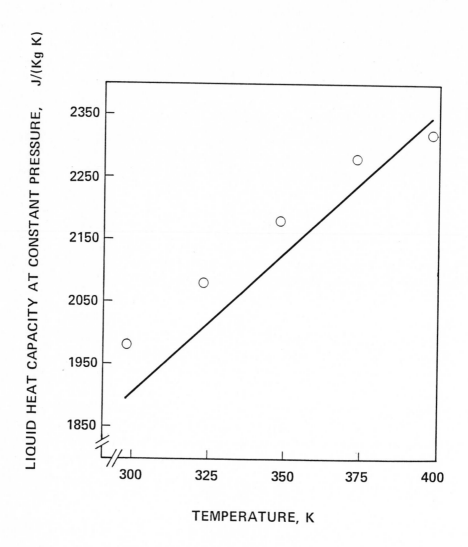

Figure VIII-2. Variation of the Heat Capacity with Temperature for Narrow Boiling, Fraction 6. ———, Group additivity estimates; O, experimental data.

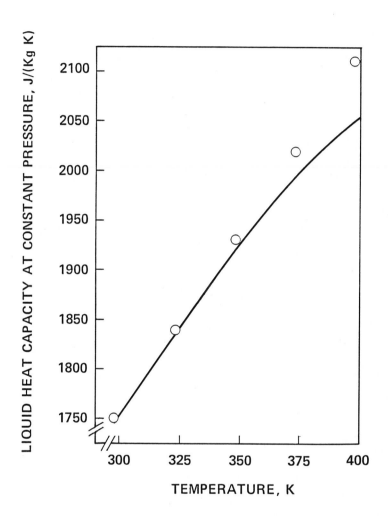

Figure VIII-3. Variation of the Heat Capacity with Temperature for Narrow Boiling, Fraction 10. ———, Group additivity estimates; O, experimental data.

176

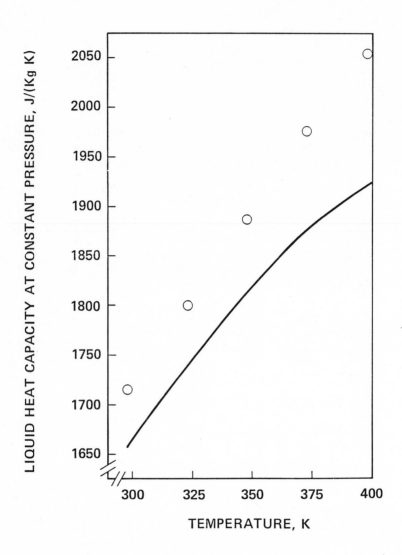

Figure VIII-4. Variation of the Heat Capacity with Temperature for Heavy
              Distillates. ———, Group additivity estimates; O, experi-
              mental data.

Table VIII-4

Estimates from Group-Contribution Methods and Observed Values
of Heat Capacity at Constant Pressure

Temperature		Heat Capacity in J/kg K		
K		Fraction #6	Fraction #10	Heavy Distillate
298	observed	1980	1750	1715
	calculated	1894	1746	1657
	Δ	-4.34%	-0.23%	-3.38%
323	observed	2080	1840	1800
	calculated	2007	1838	1739
	Δ	-3.51%	-0.11%	-3.39%
348	observed	2180	1930	1886
	calculated	2122	1920	1813
	Δ	-2.66%	-0.52%	-3.87%
373	observed	2280	2020	1975
	calculated	2235	1995	1875
	Δ	-1.97%	-1.24%	-5.06%
398	observed	2320	2110	2053
	calculated	2346	2050	1921
	Δ	+1.12%	-2.84%	-6.43%

Δ = % Deviation = 100 x (calculated-observed)/observed.

Table VIII-5

Summary of Deviations in Predicted Liquid Heat Capacities for Coal Liquids

Source	% Deviation[a]					
	Fraction #6			Fraction #10		
	av	bias	max	av	bias	max
Group contribution	2.72	-2.28	-4.34	0.99	-0.99	-2.84
Starling et al.	8.48	+8.48	+11.65	19.49	-11.44	-57.17
Gulf Graphical Correlation	5.05	-5.05	-6.32	7.15	+7.15	+9.37
Original Watson-Nelson Correlation	18.06	-18.06	-19.16	9.74	-9.74	-10.39
Modified Watson-Nelson Correlation	13.10	-13.10	-14.14	1.26	-1.26	-1.60

(a) % Deviation = 100 x (calculated-experimental)/experimental.

$$av = [\sum_i |\%dev_i|]/N$$

$$bias = [\sum_i \%dev_i]/N$$

N = number of data points.

Table VIII-6
Calculated and Observed Heat Capacities for Narrow-boiling Fraction #10
Using the Liquid- phase and Gas-phase Group Contribution Values

Temperature (K)	Heat Capacity (Calculated) (J/kg K)		Deviation[a] (%)	
	Liquid	Gas	Liquid	Gas
298	1746	1368	-0.2	-21.8
323	1838	1454	-0.1	-21.0
348	1920	1547	-0.5	-19.8
373	1995	1640	-1.2	-18.8
398	2050	1733	-2.8	-17.8
423	2080	1814	-5.9	-17.9
448	2081	1902	-9.9	-17.7
473	2081	1902	-9.9	-17.7
473	2048	1977	-15.4	-18.3
498	1972	2059	-22.4	-18.9
523	1934	2128	-27.6	-20.3

(a) % Dev = 100 x [(calculated-observed)/observed]

Table VIII-7

Linearized Group Contributions to Liquid Heat Capacity

Functional Group Number	A*	B*
1	.300	.314
2	.444	.461
3	.588	.598
4	.632	.625
5	.714	.693
6	.144	.137
7	.0212	.083
8	.0032	.163
9	-.0600	-.031
10	.0417	.166
11	.0696	.046
12	.0667	.487
13	.270	.269
14	.200	.004
15	.0715	.112
16	.122	.000
17	.132	.227
18	.290	.253
19	.130	.098
20	.0892	.038
21	.0212	.083
22	.0697	.046
23	.0805	.086
24	.0101	.100
25	.0502	.069

$^{*}C_p$ = A+B T/1000 where T is in °F and C_p is in BTU/(lb °F) or (cal/g °C).

and the ^{1}H NMR, through estimates of aromatic ring concentrations, gives an indication of the extent of π-π bonding. These comprise the major non-idealities for coal liquids, so it is not surprising that they provide suffi-cient data for heat capacity estimates. Thus, group concentrations, estimated with only elemental analysis and ^{1}H NMR data, can be used to estimate coal liquid heat capacities with an accuracy of ±5% (Allen, 1986a).

VIII.A.2 Other Properties

Group contributions for the enthalpy of formation, entropy of formation and heat capacity as a function of temperature are available for almost all func-tional groups pertinent to fuel processing (Benson, 1976; Missenard, 1965; Luria and Benson, 1977). From these three properties, most thermodynamic variables can be determined at any desired temperature.

It is important to note that while the heat capacities estimated in the previous section were for the liquid phase, most of the group contributions are for ideal gas properties.

Additional thermodynamic properties that can be calculated using a group contribution approach include critical properties (Allen, 1986b) and activity coefficients (Fredenslund, 1975; Pedersen, 1984a,b,c).

VIII.B EQUATION OF STATE PARAMETERS

Structural characterizations have also been used as parameters in equations of state. Alexander (1985a,b) used average structural parameters of the type described in Chapter V in developing correlations for the constants in an equation of state. The equation of state was designed for use with crude oils and coal derived liquids and was of the form:

$$P = \frac{RT}{V-b} - \frac{a(T)}{V(V+b)}$$

where P is the total pressure, and V is the molar volume. The variable a(T) is a function of temperature and the values of the structural parameters of the fuel being modeled. The variable b is a function of the structural param-eters alone. The equation is similar to the Van der Waals equation in form, so the parameters a(T) and b have the same physical significance as those values, i.e., b is an excluded volume and a(T) is a measure of intermolecular forces.

Given a molecular weight and the values of the structural parameters describing an average molecule, it seems reasonable to expect that representa-tive excluded volumes and intermolecular force constants could be calculated. Alexander (1985b) developed this approach and compared the predicted results to both experimental data and the predictions of other correlations. The results were comparable to the predictions of other equations of state but like the other equations of state, errors on the order of 50% were observed for some samples.

REFERENCES

Alexander, G. L., Creagh, A. L. and Prausnitz, J. M., Ind.   Eng.   Chem.   Fundam., $\underline{24}$, 301 (1985a).

Alexander, G. L., Schwarz, B. J. and Prausnitz, J. M., Ind.   Eng.   Chem.   Fundam., $\underline{24}$, 311 (1985b).

Allen, D. T., Petrakis, L., Grandy, D. W., Gavalas, G. R. and Gates, B. C., Fuel, $\underline{63}$, 803 (1984).

Allen, D. T., Grandy, D. W., Jeong, K. M. and Petrakis, L.,   Ind. Eng. Chem. Process Des. Dev., $\underline{24}$, 737 (1985).

Allen, D. T., Behmanesh, N., White, C. M. and Perry, M. B., to be submitted to Fuel (1986a).

Allen, D. T. and Behmanesh, N., to be submitted to Fuel (1986b).

Benson, S. W. "Thermochemical Kinetics", 2nd Ed., Wiley, New York, 1976.

Cotterman, R. L., Bender, R. and Prausnitz, J. M., Ind.   Eng.   Chem.   Process Des. Dev., $\underline{24}$, 194 (1985).

Fredenslund, A., Jones, R. L. and Prausnitz, J. M., AIChE J., $\underline{21}$, 1096 (1975).

Gray, J. A. "Selected Physical, Chemical, and Thermodynamic Properties of Narrow Boiling Range Coal Liquids for the SRC-II Process", Report No. DOE/ET/10104-7, 1981.

Gray, J. A. and Holder, G. D. "Selected Physical, Chemical, and Thermodynamic Properties of Narrow Boiling Range Coal Liquids from the SRC-II Process, Supplemental Property Data", Report No. DOE/ET/10104-44, 1982.

Le, T. T. and Allen, D. T., Fuel, $\underline{64}$, 1754 (1985).

Luria, M. and Benson, S. W., J. Chem. Eng. Data, $\underline{22}(\underline{1})$, 90 (1977).

Missenard, F. A., Comp. Rend., $\underline{260}(\underline{21})$, 5521 (1965).

Mraw, S. C., Heldman, J. L., Hwang, S. C. and Tsonopoulos, C., Ind. Eng. Chem. Process Des. Dev., 23, 577 (1984).

Pedersen, K. A., Thomassen, P. and Fredenslund, A., Ind. Eng. Chem. Process Des. Dev., 23, 163 (1984a).

Pedersen, K. S., Thomassen, P. and Fredenslund, A., Ind. Eng. Chem. Process Des. Dev., 23, 566 (1984b).

Pedersen, K. S., Thomassen, P. and Fredenslund, A., Ind. Eng. Chem. Process Des. Dev., 23, 948 (1984c).

Reid, R. C. and San Jose, J. L., Chem. Eng., 83(4), 161, 67 (1976).

Reid, R. C., Prausnitz, J. M., Sherwood, T. K., "The Properties of Gases and Liquids", 3rd Ed., McGraw Hill, New York, 1977.

Ruzicka, V., Fredenslund, A. and Rasmussen, P., Ind. Eng. Chem. Process Des. Dev., 22, 49 (1983).

Starling, K. E., Lee, L. L. and Kumar, K. H. "Development of a Self-Consistent Thermodynamic- and Transport-Property Correlation Framework for the Coal Conversion Industry", Report No. DOE/PC/30249-T1, 1981.

Chapter IX

CHARACTERIZATION OF ASPHALTENES

We have reviewed the application of characterization techniques to whole fuels and fractionated fuels. We have considered all of the fuel fractions equally, however, one fraction in particular exerts a dominating influence on the properties of heavy fuels. The molecules that form the fraction known as "asphaltenes" are believed to form a micellar phase in heavy fuels and thus they are qualitatively different than most other fuel fractions. A qualitative picture of asphaltene colloidal structure has been given by Speight (1984) and others and is shown schematically in Figure IX-1.

These micellar structures are not detected directly, rather they are measured indirectly as a solubility fraction. Specifically, the term "asphaltenes" denotes fossil-fuel-derived material soluble in benzene or toluene but insoluble in a low-boiling n-alkane solvent, usually n-pentane. Asphaltenes occur in unprocessed heavy crudes and comprise much of the intermediate products in direct coal liquefaction processes (Bunger and Li, 1981; Ladner, 1980). Since the asphaltene content has such a dramatic effect on the chemical nature of liquid fuels, the characterization of asphaltenes may be helpful in many aspects of fuel processing. In this chapter the average chemical structures of asphaltenes from three coal liquefaction processes are deduced using an analytical scheme based on NMR spectroscopy. The structural information is then used to construct a pseudo-phase diagram for asphaltenes.

IX.A ORIGIN OF ASPHALTENES

The samples from which the asphaltene fractions were separated were the SRC-II heavy distillate described in Chapter VII, two super critical gas (SCG) extracts and the extract from a hydrogen donor solvent (HDS) liquefaction process. The heavy distillate and one of the SCG extracts were prepared from a U.S. (Powhatan No. 5) coal and the other SCG extract and the HDS extract were prepared from a U.K. (Daw Mill) bituminous coal. Thus, this collection of asphaltenes represents material generated using two different coals under both severe (SRC liquefaction) and moderate (SCG extracts) conditions.

IX.B STRUCTURE OF ASPHALTENES

A detailed discussion of the analytical data is provided by Snape (1984a). The structural analysis scheme used is essentially one of the many subtle permutations of the parameters approach discussed in Chapter VI (Herod, 1981).

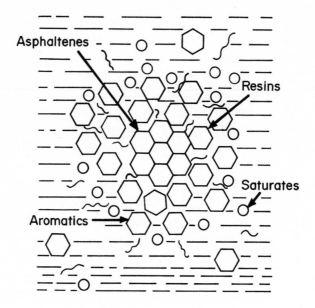

Figure IX-1. Micellar Structure of Asphaltenes (Speight)

In this formulation, numbers of atoms and groups per average molecule were calculated by combining data from elemental, molecular weight, NMR, hydroxyl and basic nitrogen analyses. These numbers were then used to calculate the numerical values of the structural parameters listed in Table IX-1. Some idea of the types of structures that can be represented by this data are given by considering equimolar mixtures of the molecules shown in Fig. IX-2. This mixture would result in average parameters that closely fit the values of the structural parameters determined for the asphaltenes. These molecules, of course, only represent an average of size and chemical type for the extremely large number of species present in the asphaltenes.

The structural parameters of Table IX-1 and average structures indicate that, as well as being more aromatic, the heavy distillate and HDS extract asphaltenes contain more condensed and less substituted aromatic nuclei (lower DC and σ values) than the SCG extract asphaltenes from both the U.S. and U.K. coals. In Fig. IX-2, only 1 and 2-ring nuclei are needed to represent the aromatic structures in the SCG extract asphaltenes compared with 2- 3- and 4-ring structures in the heavy distillate and HDS extract asphaltenes. These findings have been confirmed by polarography (Ladner and Snape, 1978) which indicated ≥3-ring structures in the heavy distillate and HDS extract asphaltenes but not in the SCG extract asphaltenes.

The combination of small alkyl and hydroaromatic ring substituents used to represent the aliphatic structures in Fig. IX-2 for the heavy distillate asphaltenes is in sharp contrast to the large naphthenic and hydroaromatic ring systems proposed by Whitehurst (1979), Farcasiu (1979) and Wooton (1978) for other SRC products. The ^{13}C NMR techniques described in previous sections can be used to resolve the conflict, however. Spin-echo methods have been used to show that quaternary aliphatic carbons, which would be anticipated in condensed naphthenic structures, are not present in these coal liquefaction products.

In a broad context, these results demonstrate that asphaltenes are not uniform in their chemical structure. They are, by definition, a solubility fraction and thus can contain a broad spectrum of molecular types. Snape and Bartle (1984b) have made this point clear by defining an asphaltene "phase diagram", shown in Figure IX-3. The axes of this "phase diagram" are average structural parameters, calculated in the manner discussed in Chapter VI. The figure indicates that asphaltenes encompass an enormous range of molecular types, from low molecular weight, hydrogen bonding structures to high molecular weight condensed ring systems.

The volume representing the asphaltene region in this diagram was originally defined by equation 1 (Snape, 1984b)

Table IX-1

Structural parameters for asphaltenes

Parameter	SRC-II heavy distillate	SCG extracts		HDS extract
		Powhatan No. 5 1	Daw Mill 2	
Aromaticity	0.81	0.72	0.76	0.83
Aliphatic H/C ratio	2.2	2.2	2.15 (2.25)[a]	2.15
No. of aliphatic substituents excluding $ArCh_2Ar$ groups (rings) count as 2 substituents)	1.8	6.1-6.6	4.5-4.9	2.5-2.7
Average no. of carbons in aliphatic substituents	1.6	1.8	1.7	1.6
No. of ring-joining groups	0.75	2.4	2.2	1.0
No. of peripheral aromatic carbons	11.4	25.5	22.7-23.1	17.0→17.2
No. of internal or bridge-head aromatic carbons	2.6	5.4-4.9	5.0-4.6	5.2-5.0
Degree of condensation of aromatic nuclei	0.62	0.71-0.73	0.70-0.72	0.64-0.65
Degree of aliphatic substitution on aromatic nuclei	0.18	0.30-0.32	0.25-0.28	0.15-0.18

a. by spin-echo ^{13}C NMR

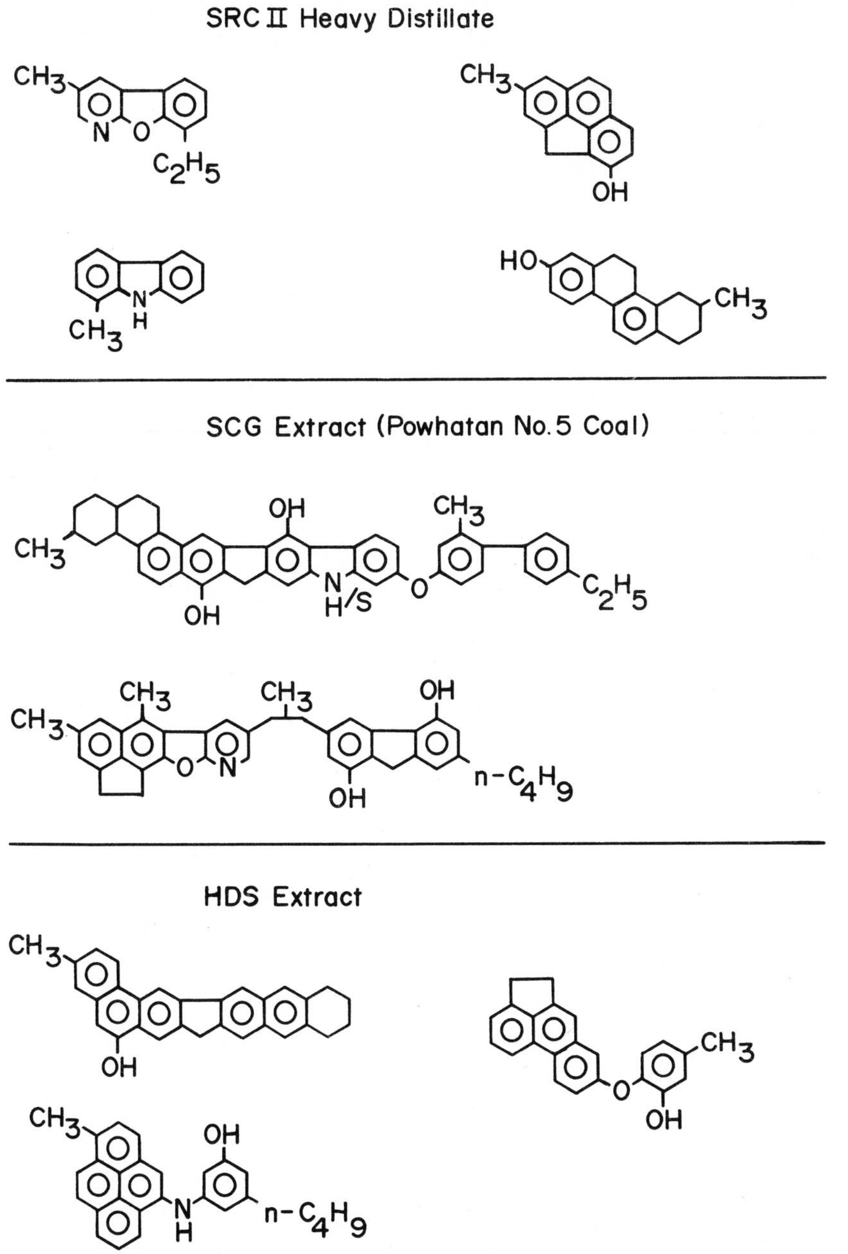

189

Figure IX-2.   Average Molecular Structures for 3 Coal Derived Asphaltenes

190

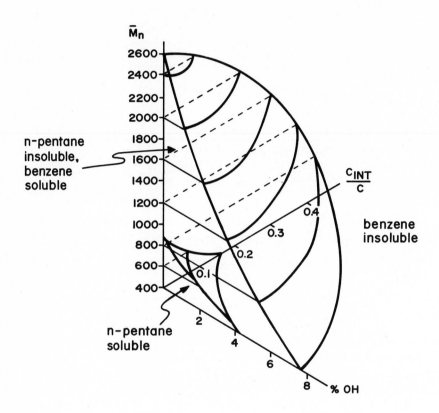

Figure IX-3. Asphaltene "Phase Diagram"

$$0.85 \leq \log_{10}\left(\frac{M_n}{200}\right)^{0.75} + 0.1(\%\text{acidic OH}) + 1.5C_{INT}/C \leq 1.2 \qquad (1)$$

where M_n is the number average molecular weight and C_{INT}/C is the ratio of internal (bridgehead) aromatic carbon to total carbon. Recently Lin (1986) has proposed a modified form of this equation that accounts for the effects of nitrogen groups that participate in hydrogen bonds

$$150 \leq (M_n^{2/3}) + 12.4(\%N + \%\text{acidic OH})$$

$$+ 66.7\left\{\begin{matrix}\text{fraction of carbon}\\ \text{that is atomatic}\end{matrix}\right\} \leq 300 \qquad (2)$$

IX.C CONCLUSIONS

1. This section has shown that, compared with asphaltenes prepared by the relatively mild SCG extraction process, those in an HDS extract and an SRC-II heavy distillate and, indeed, in other pilot plant SRC products contain:

   a. More aromatic groups;

   b. more condensed aromatic nuclei which on average can be largely represented by 2- to 4-ring structures;

   c. less oxygen and sulphur;

   d. smaller aliphatic substituents.

   Thus, compared with SCG extracts, HDS and SRC products have undergone much more severe dehydrogenation and cracking reactions.

2. The distribution of aliphatic substituents in the SRC-II heavy distillate asphaltenes is considerably different from those suggested by other workers for SRC products in that they contain fewer condensed naphthenic ring systems. However, confidence can be placed in the distribution deduced here due to the improved assignment and measurement of aliphatic carbon peaks in ^{13}C NMR spectra.

3. These points serve to illustrate the fact that asphaltenes are a solubility fraction and are not uniform in their functionality. Perhaps the most logical structural definition of asphaltenes is the phase diagram of Figure IX-3.

REFERENCES

Bunger, J. W. and Li, N.C. (Eds.), Chemistry of Asphaltenes, Am. Chem. Soc. Adv. in Chem. Series No. 195 (1981).

Farcasiu, M., Prepr. Am. Chem. Soc., Div. Fuel Chem., $\underline{24}(\underline{1})$ 121 (1979).

Herod, A. A., Ladner, W. R. and Snape, C. E., Phi. Trans. Roy. Soc. Lond., $\underline{A300}$, 3 (1981).

Ladner, W. R. and Snape, C. E., Fuel, $\underline{57}$, 658 (1978).

Ladner, W. R., Martin, T. G., Snape, C. E. and Bartle, K. D., Prepr. Am. Chem. Soc., Div. Fuel Chem., $\underline{25}(\underline{4})$, 67 (1980).

Lin, C. and Allen, D. T., submitted to Fuel (1986).

Snape, C. E., Ladner, W. R., Petrakis, L., and Gates, B. C., Fuel Processing Technology, $\underline{8}$, 155 (1984a).

Snape, C. E., and Bartle, K. D., Fuel, $\underline{63}$, 883 (1984b).

Speight, J. G. in "Characterization of Heavy Crude Oils and Petroleum Residues", Technip, Paris, 1984.

Whitehurst, D. D., Farcasiu, M. and Mitchell, T. O., EPRI report AF-1298, volumes 2 and 3 (1979).

Wooton, D. L., Coleman, W. M., Taylor, L. T. and Dorn, H. C., Fuel, $\underline{57}$, 17 (1978).

Chapter X

¹⁷O NMR APPLIED TO FOSSIL FUELS

Since carbon and hydrogen are the dominant atomic species in fuels, we have primarily focussed on the use of ^{1}H and ^{13}C NMR in fuel research. NMR of other nuclei are also potentially significant. For example, oxygen species are a significant factor in the stability and usefulness of natural and synthetic fossil fuels. Oxygen is the most abundant heteroatom in most coals and requires hydrogen consumption for its removal, a requirement which can adversely affect the economics of coal liquefaction. Oxygen bonds (for example, ethers), on the other hand, may be the "weak link" in the depolymerization of coal. Thus, the unequivocal identification and quantitative determination of oxygen functionalities could greatly aid in the design of processes for producing and upgrading synthetic fuels in a cost-effective manner (Gorbarty, 1979). Nuclear magnetic resonance of naturally abundant ^{17}O can be used to determine oxygen functionalities in synthetic fuels.

^{17}O has found only limited use in the study of natural substances because of its low natural abundance (0.037%) and the broad lines encountered due to its quadrupolar nucleus (I = 5.2). However, with improvements in Fourier transform NMR instrument sensitivity, ^{17}O NMR may become very important as a tool for studying oxygen functionalities in natural systems previously considered to be too low in oxygen for successful study (Gerothanassis, 1982). Despite its difficulties, ^{17}O NMR has several important advantages. The chemical shift of ^{17}O has a range of ≈800 ppm, and the ^{17}O chemical shifts of many small (molecular weight <100) compounds have been reported throughout this range (Christ, 1961; Sugawara, 1979; St. Amour, 1981; Kalabin, 1982). The quadrupolar nature of the nuclei which broadens the observed lines also shortens the spin-lattice relaxation time (T_1) of the nuclei, such that very rapid pulsing is possible; more than a million transients can be recorded overnight. Determination of the relative amounts of the various oxygen functionalities can also be done, if ^{17}O spectra with sufficient signal-to-noise ratio are available and the various signals are resolved.

The samples we will use to demonstrate the utility of ^{17}O NMR are the heavy distillates produced from Powhatan No. 5 coal using the SRC-II process, and three acidic and two basic fractions that had been prepared chromatographically from the heavy distillate (Petrakis, 1983). This is the same coal derived liquid that we have used to demonstrate techniques in previous

194

sections. The oxygen content of the whole fractionated sample was; total heavy distillate, 2.3%; very weak bases, 7.2%; strong bases, 1.8%; very weak acids, 8.9%; weak acids, 9.8%; and strong acids, 17.9% (Chapter VII provides additional data on these materials).

The spectra were obtained with a Varian XL-200 spectrometer using a 14 μs pulse (90°), a 13 ms acquisition time and a 20 ms delay. Using this sequence, about 30 transients s^{-1} or 108,000 h^{-1} can be obtained, yielding a good spectrum within a few hours. Deuterated methylene chloride was used as the solvent for acid and base fractions; it was also used in all samples as the field frequency lock. From $3x10^5$ to $2.7x10^6$ transients were recorded before Fourier transformation. Figure X-1 gives the ^{17}O spectra of the very weak acids (A1), weak acids (A2) and strong acids (A3). All three of these samples were spiked with 50 μℓ each of methanol and acetone to aid in the adjustment of the spectral phasing. Methanol appears as the peak at ~-35 ppm and acetone is the prominent peak at 540 ppm. All three spectra also exhibit a peak near 0 ppm, identifiable as water. Both the weak acid (A2) and the very weak acid (A1) spectra display a large broad peak, resolvable in A1, in the -10 to 80 ppm region. This is characteristic of alcohols, phenols and ethers (Christ, 1961; Sugawara, 1979).

There seem to be some aldehyde and/or ketone functional groups in the very weak acid (Fig. X-1a), as indicated by an upfield shoulder on the acetone peak. Both the very weak and weak acids as well as their hydrotreated products have been exhaustively studied by gas chromatography/mass spectrometry methods (Katti, 1983; Li, 1986). These methods show the very weak acid to contain dimethyl benzofuranone, cyclohexylphenol isomers, ethylphenoxybenzene, methylcyclohexylphenol, trimethyldihydronaphthalenone, tetrahydronaphthol and phenylphenol. The weak acid contains tetrahydronaphthol, phenylphenol, methyltetrahydronaphthol, methylphenylphenol, cyclohexylphenols and dimethylphenoxybenzene. The ^{17}O spectrum of the strong acids (A3) shows only the prominent acetone and methanol peaks, a small peak around 0 ppm, probably due to water, and a peak at ~230 ppm. This is the appropriate spectral region for furans and the carbonyl oxygen of carboxylic acids. The chemistry of the separation procedure and the IR data indicate that carboxylic acids are the most likely candidates (Petrakis, 1983).

The ^{17}O spectrum of the total heavy distillate was also recorded. Although $2.7x10^6$ scans were used, the spectrum is noisy. Integration measurements indicate that about 19% of the oxygen is in the alcohol, phenol or ether region of -50 to 100 ppm. Phenols are most likely. About 44% of the oxygen is in the carboxylic acid (C = 0) and furan region at 150-280 ppm. The chemical shift range of 310-380 ppm contains ~14% of the oxygen. Esters are the

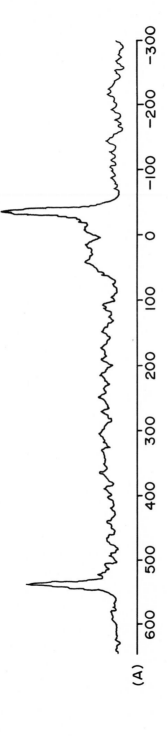

Figure X-1a. ^{17}O NMR Spectrum of Very Weak Acid Fraction (A1)

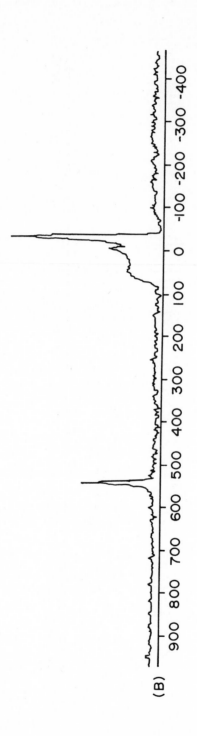

Figure X-1b. ^{17}O NMR Spectrum of Weak Acid Fraction (A2)

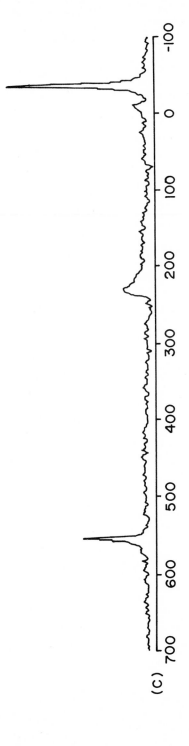

(C)

Figure X-1c. ^{17}O NMR Spectrum of Strong Acid Fraction (A3)

only reported compounds having ^{17}O NMR signals in this region (Christ, 1961; Sugawara, 1979); however, the ^{17}O NMR shifts of many of the oxygen-containing aromatic compounds that may be found in coal liquids, such as quinones and aromatic acids, are unknown. The carbonyl region, due to ketones and aldehydes (480-550 ppm), (Christ, 1961; St. Amour, 1981) contains about 23% of the oxygen. These determinations are speculative at this stage and better spectra are needed for more reliable quantitative measurements.

As Fig. X-1a shows, a very favorable situation for quantitative analysis exists in the ^{17}O spectrum of the very weak acid (A1). From the spectrum integration, about 14% of the oxygen is in the ketone or aldehyde region of 500-550 ppm, 29% is in the phenol, alcohol and ether region of -10 to 80 ppm and the remainder is in the 100-500 ppm region. If a sufficient catalog of ^{17}O chemical shifts of coal-like compounds existed, one might be able to split this region into oxygen functionality sections. Overall, the integration values for the coal liquid signals and the methanol and acetone signals are consistent with the 50 $\mu\ell$ of each pure compound and a 1 g sample of coal liquid acid containing about 9% oxygen by weight.

With materials of low oxygen content (~1%), as much as 16 h of spectrometer time may be necessary to obtain useful spectra, incurring an expense that may discourage ^{17}O NMR analysis of large numbers of routine samples. Since ^{17}O is specific to one atom and the signal area is directly proportional to the concentration of the functional group, it offers an excellent method of quantitative analysis. Thus, ^{17}O NMR can be a very powerful tool for the investigation of oxygen functionalities in the complex mixtures present in synthetic fuels.

REFERENCES

Christ, H. A., Diel, P., Schneider, H. R. and Dahn, H., Helv. Chim. Acta, 44, 865 (1961).

Crandall, J. K. and Centeno, M. A., J. Org. Chem., 44, 1183 (1979).

Gerothanassis, I. P., Lauterwein, J. and Sheppard, N., J. Mag. Resonance, 48, 431 (1982).

Gorbarty,M. L., et al., Science, 206, 1029 (1979).

Kalabin, G. A., Jushnarev, D. F., Valeyev, R. B., Trofimov, B. A. and Fedotov, M. A., Org. Mag. Resonance, 18, 1 (1982).

Katti, S., Westerman, D. W. B., Gates, B. C., Petrakis, L., Grandy, D. W. and Youngless, T., Proc. 1983 Int. Conf. Coal Sci., Pittsburgh (1983).

Li, C. L., Xu, Z. R., Gates, B. C. and Petrakis, L., IEC Process Design Dev. (submitted).

Petrakis, L., Young, D.C., Roberto, R. G. and Gates, B. C., IEC Process Design Dev., 22, 292 (1983).

St. Amour, T. E., Burgar, M. I., Valentine, B. and Flat, D. J., J. Am. Chem. Soc., 103, 1128 (1981).

Sugawara, T., Kawada, Y., Katoh, M. and Iwamura, H., Bull. Chem. Soc. Japan, 52, 3391 (1979).

APPENDIX 1

FORTRAN CODE FOR CALCULATING AVERAGE STRUCTURAL PARAMETERS

    Chapter V reviewed methods for using NMR data to calculate average struc-
tural parameters. Applying most of these methods is a straightforward process, however, determining average parameters using ^{1}H NMR method 3 involves a
series of iterative calculations. Listed in Table A1-1 is a FORTRAN program
which performs a revised version of the method 3 calculations.

    The program requires 8 input variables.

1.   HMONO, the fraction of protons in monoaromatics.

2.   HDI, the fraction of protons in diaromatics.

3.   HTRI, the fraction of protons in triaromatics.

4.   HALPH1, the fraction of alpha hydrogen in CH_2 groups.

5.   HALPH2, the fraction of methyl alpha hydrogen.

6.   HBETA, the fraction of beta hydrogen.

7.   HGAMMA, the fraction of methyl hydrogen gamma or farther from an
    aromatic ring.

8.   HBETAN, the fraction of beta naphtenic hydrogen.

The variables are evaluated by integrating the ^{1}H NMR spectrum over the bands
listed in Table A1-2.

    Comparing the 8 input variables with the description of method 3 in Chapter
V reveals a number of differences. The reason for the differences is that the
program has extended the method 3 calculations to permit the inclusion of
triaromatics and to permit differentiation between alpha methylene (CH_2)
groups and alpha methyl groups. The original method 3 calculations can be
performed by using the variable definitions listed in Table A1-3.

Table A1-1

FORTRAN Code for Calculating Average Structural Parameters

```
C PROGRAM TO DETERMINE AVERAGE MOLECULAR PARAMETERS FROM NMR DATA
C
      DIMENSION AA(5),BB(3),CC(4),DD(3),EE(5),FF(5),GG(5),HH(5)
      DIMENSION A(20)
C INPUT:HMONO HDI HTRI HALPH1 HALFH2 HEETA HGAMMA HBETN
C
C AA=SPONSOR OR SUBMITTOR
C BB=SAMPLE DESCRIPTION OR REFERENCE NUMBER
C CC=PROJECT NUMBER
C DD=SAMPLE NUMBER
C EE=DATE RECEIVED
C FF=DATE REPORTED
C GG=ANALYZER
C HH=APPROVER
C  A=IBM NUMBER OR SAMPLE NUMBER
C INPUT:HMONO(2*AREA 6.95-6.0 PPM),HDI(6.0-8.15 PPM-HMONO)
C INPUT:HTRI(8.15-END OF AROM AREA),HALPH1(ALPHA CH2S 4.0-2.6)
C INPUT:HALPH2(ALPHA CH3S 2.6-1.9 PPM),HBETA(1.0-1.9 PPM)
C INPUT:HGAMMA(1.0-0.0 PPM),HBETN(NAPHTHENIC 1.9-1.65 PPM)
  999 READ(5,1,END=100)(A(I),I=1,20),HMONO,HDI,HTRI,HALP1,HALPH2,
     1HBETA,HGAMMA,HBETN
1     FORMAT(20A4,/8F10.5)
      READ(5,1111) AA,BB,CC,DD,EE,GG,HH,II
 1111 FORMAT(5A4,3A4/4A4,3A4,5A4/5A4,5A4,I10)
      WRITE(6,2)(A(I),I=1,20),HMONO,HDI,HTRI,HALPH1,HALPH2,HBETA,
     1HGAMMA,HBETN
2     FORMAT(1H1,5X,20A4,/,10X,8F10.5)
      HALPHA=HALPH1+HALPH2
      HBETN=3.0*HBETN/2.0
C PER CENT MONO AND DIAROMATIC MOLECULES
      HA=HMONO+HDI+HTRI
      TS=0.0
      RS=0.0
      DO 200 IN=1,10
      PMONO=HMONO/(HMONO+HDI*(6.0-RS)/(8.0-RS))
      AINT=(8.0-2.0*PMONO)/(HMONO+HDI+HALPH1/2.0+HALPH2/3.0
      RS=AINT*HALPH1/2.0+HALPH2/3.0)
      TEST=ABS(RS-TS)
      IF(TEST.LT.0.005) GO TO 202
      TS=RS
  200 CONTINUE
      WRITE(6,201)
  201 FORMAT(10X,"ITERATION DID NOT CONVERGE IN TEN PASSES')
  202 CONTINUE
      PMONOP=PMONC*100.0
      PDI=1.0-PMONO
      PTRI=0.0
      IF(HDI.LT.0.05) GO TO 203
      PTRI=PDI*(8.0-RS)*HTRI/(2.0*HDI)
      PDI=PDI-PTRI
  203 CONTINUE
      PDIP=PDI*100.0
```

Table A1-1 (continued)

```
      PTRIP=PTRI*100.0
C NUMBER MONO AND DIAROMATICS
      PPMONO=PMONO*6.0
      PPDI=PDI*10.0
      PPTRI=PTRI*14.0
C TOTAL NUMBER CA= NUMBER MONO PLUS NUMBER DIAROMATICS(WITH BRIDGES)
      TOTLCA=PPMONO+PPDI+PPTRI
C NUMBER OF BRIDGEHEADS= PER CENT DIAROMATICS * 2,TRI*4
      BRIDG=PDI*2.0+PTRI*4.0
C TOTAL NUMBER C1 = TOTLCA - BRIDG
      TOTLC1=TOTLCA-BRIDG
C TOTAL OF OBSERVED CARBONS(BRIDGES AND SUBST. AROM CARBONS EXCLUDED)
      HBETA=HBETA-HBETN
      SUM=HA+HALPH1/2.0+HALPH2/3.0+HBETN/1.5+HBETA/2.0+HGAMMA/3.0
C PROPORTIONS OF DIFFERENT CARBON ATOMS TO TOTAL OBSERVED CARBONS
      AHA=HA/SUM
      AHALFA=(HALPH1/2.0+HALPH2/3.0/SUM
      AHALF1=HALPH1/2.0/SUM
      AHALF2=HALPH2/3.0/SUM
      AHBETA=HBETA/(2.0*SUM)
      AHGAMA=HGAMMA/(3.0*SUM)
      AHBETA=HBETN/(1.5*SUM)
C INTENSITY UNITS/CARBON
      CUNITS=(AHA+AHALFA)/TOTLC1
C NUMBERS OF DIFFERENT CARBONS
      EAR=TOTLCA
      EALFA=AHALFA/CUNITS
      EALF1=AHALF1/CUNITS
      EALF2=AHALF2/CUNITS
      EBETA=AHBETA/CUNITS
      EGAMMA=AHGAMA/CUNITS
      EBETN=AHBETN/CUNITS
C NUMBER OF CARBONS/ AVERAGE MOLECULE LESS BRIDGEHEADS
      EBETA=EBETA+EBETN
      TOTLC=EAR+EALFA+EBETA+EGAMMA
C AROMATICITY
      FA3=FAR/TOTLC
C PERCENT SUBST. OF ALKYLS ON PERIPH. SUB. CARBONS (PERCENT AS)
      AASM=RS*100.0/6.0
      AASD=RS*100.0/8.0
      AAST=RS*100.0/10.0
      AAS=AASM*PMONO+AASD*PDI+AAST*PTRI
      AS=AAS/100.0
C PROTON CALCULATIONS FOLLOW
C TOTAL SUM AND INDIVIDUAL PROPORTIONS
      SUMH=    HA    +HALPHA+HBETA+HGAMMA+HBETN
      HAA=HA/SUMH
      HALFAA=HALPHA/SUMH
      HBETAA=HBETA/SUMH
      HGAMAA=HGAMMA/SUMH
      HBETNA=HBETN/SUMH
C INTENSITY UNITS/PROTCN

      HUNITS=HAA/((1.00-AS)*TOTLC1)
C NUMBERS OF DIFFERENT PROTONS
      FAR=(1.0-AS)*TOTLC1
      FALFA=HALFAA/HUNITS
      FBETA=HBETAA/HUNITS
      FGAMMA=HGAMAA/HUNITS
      FBETN=HBETNA/HUNITS
C NUMBER OF PROTONS / AVERAGE MOLECULE
      FBETA=FBETA+FBETN
      TOTLH=FAR+FALFA+FEETA+FGAMMA
```

Table A1-1 (continued)

```
C AVERAGE MOLECULAR FORMULA= 'TOTLC;TOTLH'
C AVERAGE MOLECULAR WEIGHT
      AMW=TOTLC*12.011+TOTLH*1.008
C AVERAGE PERCENT CARBON
      PCARB=TOTLC*12.011/AMW
C AVERAGE PERCENT HYDROGEN
      PHYDR=TCTLH*1.008/AMW
C AVERAGE C/H RATIO OF ALIPHATIC PORTION OF THE MOLECULE
      F=(EALFA+EBETA+EGAMMA)*12.01/(FALFA+FBETA+FGAMMA)*1.008
C AVERAGE NUMBER OF AROMATIC RINGS
      RA=BRIDG/2.0+1.0
C PERCENT SUBST. OF AROMATIC CARBON ATOMS
      PC1=TOTLC1/TOTLC
C PERCENT ALKYL CARBON
      PCS=(EALFA+EBETA+EGAMMA)/TOTLC*100.0
C AVERAGE NUMBER OF CARBONS/ SUBSTITUENT
      S=(EALFA+EBETA+EGAMMA)/EALFA
C NUMBER OF NAPHTHENE RINGS/AVERAGE MOLECULE
      RN=FBETN/3.0
C NUMBER OF NAPHTHENE RINGS/ALKYL GROUP
      R=RN/RS
C PERCENT NONBRIDGE AROMATIC RING CARBONS
      PNB= (TOTLC1/TOTLC)*100.0
C PERCENT NAPHTHENIC CARBON
      PNAC= (RN*3.5/TOTLC)*100.0
      WRITE(6,1000)
 1000 FORMAT(1H1,/,T31,'AVERAGE COMPOSITIONAL PARAMETER
     2S OBTAINED',/,T34,'FROM NUCLEAR MAGNETIC RESONANCE TO',/,35,'CHARAC
     3TERIZE PETROLEUM FRACTIONS')
      WRITE(6,1001)AA,BB,CC,DD,EE,FF,GG,HH,A
 1001 FORMAT(1H0//,T10,'SUBMITTED BY: ',5A4,T46,'REFERENCE NO: ',3A4,4X,
     1'PROJ. NO.: ', 4A4,/T10,'SAMPLE NO.: ',3A4,T46,'RECEIVED: ',5A4,
     2'REPORTED: ',5A4/T10,'ANALYZED BY: ',5A4,T46,'APPROVED BY: ',5A4,
     3/T10,'SAMPLE DESC: ',20A4)
      WRITE(6,1002)
 1002 FORMAT(1H0,////,T46,'REPORT OF ANALYSIS'//,2X,' ')
      WRITE(6,1003)FA34,PMONOP
 1003 FORMAT(1H0,T10,'AROMATICITY:',T37,F12.2,T53,'% MONOAROMATICS:',T91
     1,F6.1)
      WRITE(6,1004)RA,PDIP
 1004 FORMAT(1HO,T10,'AROMATIC RINGS/MOLECULE:',T15.1,T53,'% DIAROMATICS
     1:',T91,F6.1)
      WRITE(6,1005)TOTLCA,PTRIP
 1005 FORMAT(1H0,T10,'AROMATIC RING CARBONS/MOLECULE:',F8.1,T53,'% TRIAR
     1OMATICS:',T91,F6.1)
      WRITE(6,1006)TOTLC,TOTLH,AMW
 1006 FORMAT(1H0,T10,'AVERAGE MOLECULAR FORMULA:',4X,'C',F4.1,' H',F4.1,
     1T53,'AVERAGE MOLECULAR WEIGHT:',T90,F7.1)
      WRITE(6,1007)PCS,TOTLC1
 1007 FORMAT(1H0,T10,'% SATURATE CARBON:',F21.1,T53,'NONBRIDGE AROMATIC
     1CARBONS/MOLECULE:',F8.2)
      WRITE(6,1008)RS,AAS
 1008 FORMAT(1H0,T10,'ALKYL SUBSTITUENTS/MOLECULE:',F11.1,T53,'% SUBST O
     1F NONBRIDGE AROM CARBONS:',F10.1)
      WRITE(6,1009)S,RN
 1009 FORMAT(1H0,T10,'CARBONS/ALKYL SUBSTITUENT:',F13.1,T53,'NAPHTHENE R
     1INGS/MOLECULE:',F19.1)
      WRITE(6,1010)
 1010 FORMAT(/////,T10,'NOTE:  AVG MOL WT IS BASED ONLY ON C AND H AND',
     1/,T17,'DOES NOT ACCOUNT FOR O,S,N,ETC.')
      GO TO 999
  100 CALL EXIT
      STOP
```

Table A1-2

Chemical Shift Assignments for the Revised Method 3 Calculations

	Variable	^{1}H NMR Chemical Shift Range (ppm)
1.	HMONO	twice the integrated intensity of the region $6.0 \leq \delta \leq 6.95$
2.	HDI	$6.0 \leq \delta \leq 8.15$ minus the value of HMONO
3.	HTRI	$8.15 \leq \delta \leq$ end of aromatic region
4.	HALPH1	$2.6 \leq \delta \leq 4.0$
5.	HALPH2	$1.9 \leq \delta \leq 2.6$
6.	HBETA	$1.0 \leq \delta \leq 1.9$
7.	HGAMMA	$1.0 \leq \delta \leq 0.0$
8.	HBETAN	$1.9 \leq \delta \leq 1.65$

Table A1-3

Chemical Shift Assignments for the Original Method 3 Calculations

	Variable	^{1}H NMR Chemical Shift Range (ppm)
1.	HMONO	$6.0 \leq \delta \leq 7.05$
2.	HDI	$7.05 \leq \delta \leq$ end of aromatic region
3.	HTRI	0
4.	HALPH1	$1.9 \leq \delta \leq 4.0$
5.	HALPH2	0
6.	HBETA	$1.0 \leq \delta \leq 1.9$
7.	HGAMMA	$1.0 \leq \delta \leq 0.0$
8.	HBETAN	$1.9 \leq \delta \leq 1.65$

APPENDIX 2

FORTRAN CODE FOR ESTIMATING FUNCTIONAL GROUP CONCENTRATIONS IN HEAVY OILS

   Chapter VII describes a method for characterizing fuels in terms of functional group concentrations. The principles of the method were discussed but
the details of the calculation procedure were not given. This Appendix gives
a detailed explanation of the calculation methods and presents a FORTRAN program called OILS for estimating functional group concentrations in heavy oils.
   OILS is user-friendly and operates interactively. The program is written
for an IBM type PC and it begins its operation by prompting the user for the
following types of data.

1.   Elemental analysis--these data are input as weight percentages.

2. ^{1}H NMR data--the <u>fraction</u> of protons present in each of the following 4
      bands is requested

      i.   Aromatic (5.0-9.0 ppm)

      ii.   Alpha (2.2-5.0 ppm)

      iii. Beta (1.0-2.2 ppm)

      iv.   Gamma (0.5-1.0 ppm)

3. ^{13}C NMR data--the <u>fraction</u> of carbon in each of the following 6 bands is
      requested.   Note that INEPT spectra, as well as conventional spectra,
      can be used as a source for these data.

      i.   Carboxyl/Carbonyl (200-160 ppm)

      ii.   Aromatic (160-60 ppm)

      iii. Carbon in CH groups (60-37 ppm)

      iv.   Carbon in CH_2 groups (37-22.5 ppm)

      v.   Methyl carbon alpha to a ring (22.5-20 ppm)

vi.   Methyl carbon not alpha to a ring (20-0 ppm)

SARA separation yields--The yields, as weight percentages, of the following 10 fractions are requested. (The fractions differ in the heteroatomic groups assumed to be present--the group assignments for each fraction are listed later in this manual.)

i.   Neutral oils

ii.   Asphaltenes

iii. Neutral resins

iv.   Very weak bases

v.   Weak bases

vi.   Strong bases

vii. Very weak acids

viii. Weak acids

ix.   Strong acids

x.   Saturates

   The only data that is absolutely necessary for the operation of the program are elemental analyses. The program will accept any combination of the 4 data types listed above and will print out the concentrations of the 17 groups shown in Figure A2-1. An important point to note, however, is that if SARA yields are not available, the heteroatomic group assignments are made as follows:

        All sulfur is in dibenzothiophene
        All nitrogen is in carbazole
        All oxygen is in ayrl ethers

General Program Description

   An extensive discussion of the theory supporting estimates of functional group concentrations is given in Chapter VII. Four basic steps are involved in the calculations done by the program.

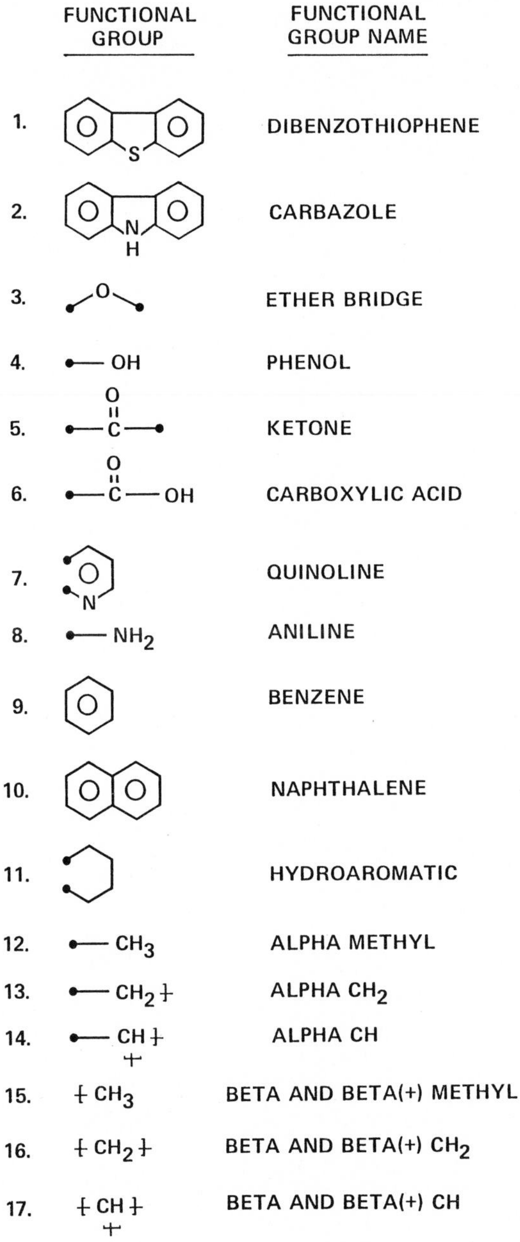

FUNCTIONAL GROUP		FUNCTIONAL GROUP NAME
1.		DIBENZOTHIOPHENE
2.		CARBAZOLE
3.		ETHER BRIDGE
4.	●— OH	PHENOL
5.	●—C(=O)—●	KETONE
6.	●—C(=O)—OH	CARBOXYLIC ACID
7.		QUINOLINE
8.	●— NH₂	ANILINE
9.		BENZENE
10.		NAPHTHALENE
11.		HYDROAROMATIC
12.	●— CH₃	ALPHA METHYL
13.	●— CH₂⊢	ALPHA CH₂
14.	●— CH⊢⊢	ALPHA CH
15.	⊢CH₃	BETA AND BETA(+) METHYL
16.	⊢CH₂⊢	BETA AND BETA(+) CH₂
17.	⊢CH⊢⊢	BETA AND BETA(+) CH

Figure A2-1. Functional Groups for OILS Program

Step 1: <u>Choose a set of functional groups</u>.

The proposed set of functional groups must be sufficient to explain the observed data yet must be concise since the number of concentrations that can be accurately estimated is limited by the amount of available data.

The program uses the 17 functional groups shown in Figure A2-1 as the primary set of groups. The hydrocarbon groups (9-17) remain the same for all combinations of data. The heteroatom concentrations are calculated only when SARA separation yields (and elemental analysis data on each fraction) are available.

Heteroatom Assignments for SARA Fractions:

The oxygen, nitrogen and sulfur functional group assignments for each SARA fraction are summarized in TAble A2-1.

Step 2: <u>Set up balance equations and determine if a solution is feasible</u>.

The program prompts the user for input data, then internally checks whether the data are consistent with the proposed set of functional groups. (It does this by first setting up a linear programming problem and then calling a subprogram to determine if a feasible solution exists.) If the data are inconsistent with the proposed groups, an error message is printed and the program terminates.

Step 3: <u>Set up the minimization problem</u>.

If the data are consistent with the chosen groups then a mathematical minimization problem is constructed. The form of the minimization problem depends on what data are available.

Data Available	Maximization or Minimization Problem
elemental analysis, ^{1}H NMR	maximize entropy subject to ^{1}H NMR and elemental analysis balance equations
elemental analysis, ^{1}H NMR, ^{13}C NMR	minimize the difference between carbon distribution predicted by ^{13}C NMR and that allowed by elemental analysis and ^{1}H NMR balance equations
elemental analysis, ^{13}C NMR	maximize entropy subject to ^{13}C NMR and elemental analysis balance equations

Table A2-1

Heteroatomic Functionalities Present in Chromatographic Fractions

SARA fraction	Oxygen	Nitrogen	Sulfur
neutral oils	phenols	carbazole	dibenzothiophene
asphaltenes	phenols	carbazole	dibenzothiophene
neutral resins	ketones, carboxylic acids	carbazole	dibenzothiophene
very weak bases	phenols	carbazole	dibenzothiophene
weak bases	phenols	pyridines	dibenzothiophene
strong bases	phenols	amines and pyridines	dibenzothiophene
very weak acids	phenols	carbazoles	dibenzothiophene
weak acids	phenols and carboxylic acids	carbazoles	dibenzothiophene
strong acids	carboxylic acids	carbazoles	dibenzothiophene

Step 4:   <u>The maximization or minimization problem is solved using a direct</u>
          <u>search</u>.

The linear programming solution found in Step 2 is used as a starting point.

An explanation of how these 4 steps are performed by the program and its subroutines is given in Table A2-2.

Sample Problem

Application of Functional Group Analysis to Heavy Oils

As a sample problem, consider the characterization of the atmospheric tower bottoms (ATB) from a refinery fed by a Mayan heavy crude oil. The analytical data available on the ATB sample are given in Tables A2-3 and A2-4.   Table A2-3 contains elemental analysis and ^{1}H NMR data. Table A2-4 gives ^{13}C NMR data.

The OILS program yielded the concentration estimates shown in the sample output (Table A2-5). The dominant features of the concentration estimates are the large amounts of long chain aliphatics and the high extent of branching in the chains.   Hydroaromatic groups and aromatic rings with no heteroatoms appear to be present in low concentration.

In the output, the values "Number of Function Evaluations = 2751" and "Minimum = .025" are reported.   The number of functional evaluations is a measure of the performance of the optimization routine. The "minimum" is a measure of the agreement between the ^{13}C NMR data and the ^{1}H NMR/elemental analysis data. The small value for the minimum indicates good agreement. These details are discussed more completely in Chapter VII.

Table A2-2

Organization of the OILS Code

Program or Subprogram Name and Hierarchy	Purpose
Main (calls EAIN, HNMRIN, CNMRIN,CONC,CONC1, SARAFG,SIMPLX)	For whole oils: calls input subprograms and the appropriate concentration estimating program; prints out resulting hydrocarbon concentrations. For fractionated oils: calls subprogram SARAFG, which handles all aspects of the calculations
CNMRIN	Subprogram for interactive input of ^{13}C NMR data
CONC (calls SIMPLX)	Subprogram for estimating hydrocarbon group concentrations. CONC is called if elemental analysis and ^{1}H NMR data are available for estimating group concentrations.
CONC1 (calls SIMPLX)	Subprogram for estimating hydrocarbon group concentrations CONC1 is called if elemental analysis and ^{13}C NMR (no ^{1}H NMR) data are available for estimating group concentrations.
HNMRIN	Subprogram for interactive input of ^{1}H NMR data.
SARAFG (calls EAIN, HNMRIN, CNMRIN,CONC,CONC1, SIMPLX)	Subprogram called if SARA data are available. SARAFG calls data input programs and the appropriate concentration estimation program; SARAFG then prints out heteroatom group concentrations.
SIMPLX	Subroutine for linear programming. Determines if functional groups are consistent with available data.
EAIN	Subprogram for interactive input of elemental analysis data.

Table A2-3

^{1}H NMR and Elemental Analysis Data for the Mayan ATB

Elemental Analysis Data				
%C	%H	%O	%N	%S
84.1	10.1	1.06	0.5	4.0

^{1}H NMR Data for Mayan ATB		
Hydrogen type	Chemical shift range (ppm from TMS)	% of Hydrogen
Aromatic hydrogen	9.0-5.0	7.7
Hydrogen in CH, CH_2 and CH_3 groups alpha to an aromatic ring	5.0-1.9	8.3
Hydrogen in CH and CH_2 groups beta or farther from an aromatic ring, Hydrogen in CH_3 groups beta to an aromatic ring	1.9-1.0	62.4
Hydrogen in CH_3 groups gamma or farther from an aromatic ring	1.0-0.5	21.6

Table A2-4

^{13}C NMR Data for the Mayan ATB

^{13}C NMR Data for Mayan ATB	
Carbon type	% of Carbon
Aromatic carbon	29.6
Carbon in CH groups	14.6
Carbon in CH_2 groups	41.1
Carbon in CH_3 groups alpha to an aromatic ring	2.7
Carbon in CH_3 groups attached to hydroaromatic structures Carbon in CH_3 groups beta to an aromatic ring Carbon in CH_3 groups gamma or farther from an aromatic ring	11.9

Table A2-5

Sample Input and Output

(Bold type indicates user input)

```
A>OILS
DATA INPUT SECTION
ARE SARA SEPARATION YIELDS AVAILABLE (01=YES 0=NO)
0
INPUT DATA ON WHOLE OIL
INPUT WT% CARBON (XX.XX)
84.1
INPUT WT% HYDROGEN (XX.XX)
10.1
INPUT WT% OXYGEN (XX.XX)
1.06
INPUT WT% NITROGEN (XX.XX)
0.5
INPUT WT% SULFUR (XX.XX)
4.
THE FOLLOWING VALUES HAVE BEEN ENTERED
%C=  84.10
%H=  10.10
%O=   1.06
%N=    .50
%S=   4.00
SUM= 99.76
OK? (01=YES, 0=NO, -1=EXIT)
1
ARE HNMR SPECTRA AVAILABLE?  (01=YES, 0=NO)
1
THE PROTON NMR SPECTRUM SHOULD BE DIVIDED INTO THE
FOLLOWING 4 BANDS:
GAMMA (0.5-1.0 PPM)
BETA (1.0-2.2 PPM)
ALPHA (2.2-5.0 PPM)
AROMATIC (5.0-9.0 PPM)
ENTER AROMATIC HYDROGEN FRACTION (0<F<1)
0.077
ENTER ALPHA HYDROGEN FRACTION (0<F<1)
0.083
ENTER BETA HYDROGEN FRACTION (0<F<1)
0.624
ENTER GAMMA HYDROGEN FRACTION (0<F<1)
0.216
THE FOLLOWING VALUES HAVE BEEN ENTERED:
AROMATIC FRACTION= .77000E-01
ALPHA FRACTION= .83000E-01
BETA FRACTION= .62400E+00
GAMMA FRACTION= .21600E+00
SUM= .10000E+01
OK?  (01=YES, 0=NO, -1=EXIT)
1
ARE CARBON SPECTRA AVAILABLE? (01=YES 0=NO)
1
THE CARBON NMR SPECTRUM SHOULD BE DIVIDED INTO THE
FOLLOWING BANDS:
200-160 PPM = CARBOXL AND CARBONYL
160-60  PPM = AROMATIC CARBON
```

Table A2-5 (continued)

60-37   PPM = CARBON IN CH
37-22.5 PPM = CH2 CARBON
22.5-20 PPM = METHYL CARBON ALPHA TO AN AROMATIC RING
20-0   PPM = METHYL CARBON NOT ALPHA TO AN AROMATIC RING
NOTE: THE CH, CH2 AND CH3 FRACTIONS MAY BE DETERMINED FROM
INEPT SPECTRA; INPUT INEPT DATA IN THE SAME WAY
ENTER CARBOXYL/CARBONYL CARBON FRACTION (0<F<1)
1.
ENTER AROMATIC CARBON FRACTION (0<F<1)
0.296
ENTER CH CARBON FRACTION (0<F<1)
0.146
ENTER CH2 CARBON FRACTION (0<F<1)
0.411
ENTER ALPHA METHYL CARBON FRACTION (0<F<1)
,027
ENTER NON-ALPHA METHYL CARBON FRACTION (0<F<1)
0.119
THE FOLLOWING VALUES HAVE BEEN ENTERED
CARBOXYL/CARBONYL FRACTION   .000
AROMATIC CARBON FRACTION   .296
CH CARBON FRACTION   .146
CH2 CARBON FRACTION   .411
ALPHA METHYL CARBON FRACTION .000
GAMMA METHYL CARBON FRACTION .119
SUM .972
OK? (01=YES, 0=NO, -1=EXIT)
0

THE CARBON NMR SPECTRUM SHOULD BE DIVIDED INTO THE
FOLLOWING BANDS:
299-160 PPM = CARBOXYL AND CARBONYL
160-60 PPM = AROMATIC CARBON
60-37 PPM   = CARBON IN CH
37-22.5 PPM = CH2 CARBON
22.5-20 PPM = METHYL CARBON ALPHA TO AN AROMATIC RING
20-0 PPM   = METHYL CARBON NOT ALPHA TO AN AROMATIC RING
NOTE: THE CH, CH2 AND CH3 FRACTIONS MAY BE DETERMINED FROM
INEPT SPECTRA; INPUT INEPT DATA IN THE SAME WAY
ENTER CARBOXYL/CARBONYL CARBON FRACTION (0<F<1)
0.
ENTER AROMATIC CARBON FRACTION (0<F<1)
0.296
ENTER CH CARBON FRACTION (0<F<1)
0.146
ENTER CH2 CARBON FRACTION (0<F<1)
0.411
ENTER ALPHA METHYL CARBON FRACTION (0<F<1)
0.027
ENTER NON-ALPHA METHYL CARBON FRACTION (0<F<1)
0.119
THE FOLLOWING VALUES HAVE BEEN ENTERED
CARBOXYL/CARBONYL FRACTION   .000
AROMATIC CARBON FRACTION   .296
CH CARBON FRACTION   .146
CH2 CARBON FRACTION   .411

Table A2-5 (continued)

```
ALPHA METHYL CARBON FRACTION   .027
GAMMA METHYL CARBON FRACTION   .119
SUM   .999
OK?  (01=YES, 0=NO, -1=EXIT)
1
```

```
NUMBER OF FUNCTION EVALUATIONS = 2751   MINIMUM= .24995E-01
HYDROCARBON GROUP CONCENTRATIONS:
    BENZENE=  .950E-03
    NAPHTHALENE= .249E-01
    ALPHA METHYL= .622E-01
    HYDROAROMATIC= .135E-03
    BETA CH2=  .192E+01
    BETA CH= .452E+00
    BETA CH3= .727E+00
    ALPHA CH2= .490E-03
    ALPHA CH= .650E+C0
    PHENOL= .622E-01
    CARBAZOLE= .375E-01
    DIBENZOTHIOPHENE= .125E+00
Stop - Program terminated.
```

Table A2-6. FORTRAN Code for OILS

```
C      THE OBJECT OF THIS PROGRAM IS TO CHARACTERIZE COMPLEX HEAVY
C      AND SHALE OIL MIXTURES BY CALCULATING THE CONCENTRATIONS OF
C      THE FUNCTIONAL GROUPS THAT MAKE UP THE MIXTURES. TO DO THIS
C      THE PROGRAM UTILIZES DATA FROM A VARIETY OF SOURCES.
C      THE CASES LISTED BELOW ARE THE COMBINATIONS OF DATA WHICH
C      THE PROGRAM CAN PRESENTLY HANDLE.
C
C      CLASS 1: SARA DATA AVAILABLE
C              MODE 0: ONLY ELEMENTAL ANALYSIS DATA AVAILABLE
C              MODE 1: ELEMENTAL ANALYSIS AND CARBON 13 NMR DATA
C              MODE 2: ELEMENTAL ANALYSIS AND PROTON NMR DATA
C              MODE 3: ELEMENTAL ANALYSIS, CARBON NMR AND PROTON NMR DATA
C
C      CLASS 2: SARA DATA NOT AVAILABLE
C              MODE 1: ELEMENTAL ANALYSIS AND CARBON 13 NMR DATA
C              MODE 2: ELEMENTAL ANALYSIS AND PROTON NMR DATA
C              MODE 3: ELEMENTAL ANALYSIS, CARBON NMR AND PROTON NMR DATA
C
C
       REAL EA(5),HNMR(4),CNMR(6),F(10),MID(9)
C
C      NOMENCLATURE
C
C      EA(5)     CONTAINS THE ELEMENTAL COMPOSITION DATA (C,H,N,O,S)
C                FOR THE SARA FRACTION CURRENTLY BEING CONSIDERED
C      F(10)     CONCENTRATIONS OF ATOMIC GROUPS; 1-4 ARE THE VARIOUS
C                H ATOM CONCS., 5-10 ARE THE VARIOUS C ATOM CONCS.
C      CNMR(6)   FRACTIONS OF C IN THE 6 BONDING ENVIRONMENTS
C      HNMR(4)   FRACTIONS OF H IN THE 4 BONDING ENVIRONMENTS
C      MID(9)    GROUP CONCENTRATIONS FOR HYDROCARBON GROUPS
C
C
C
C      DATA INPUT SECTION: BEGIN BY DEFINING WHAT DATA ARE AVAILABLE
C
C***************************************************************
       IIN=10
       IOUT=12
C***************************************************************
     5 WRITE (*,6000)
       READ(*,4000)ISARA
       IOK=0
  6000 FORMAT(' DATA INPUT SECTION',/,' ARE SARA SEPARATION YIELDS',
      2  ' AVAILABLE? (01=YES 0=NO')
  4000 FORMAT(I2)
       IF(ISARA .EQ. 0) GOTO 10
       CALL SARAFG(IIN,IOUT)
     7 CONTINUE
       WRITE(*,6010)
  6010 FORMAT(' DO YOU WISH TO RUN ANOTHER SAMPLE? (01=YES, 0=NO)')
       READ(*,4000)ISTART
       IF(ISTART .EQ. 1) GOTO 5
       STOP
C
C      SAMPLE NOT FRACTIONATED-DETERMINE HYDROCARBON CONCENTRATIONS FOR
C        THE WHOLE OIL
    10 CONTINUE
```

Table A2-6 (continued)

```
       DO 12  J=1,4
   12 HNMR(J)=0.
       DO 13  J=1,6
   13 CNMR(J)=0.
       WRITE(*,6015)
 6015 FORMAT(' INPUT DATA ON WHOLE OIL')
       CALL EAIN(IIN,IOUT,IOK,EA)
       IF(IOK .EQ. -1) GOTO 7
       WRITE(*,6017)
 6017 FORMAT(' ARE HNMR SPECTRA AVAILABLE? (01=YES, 0=NO)')
       READ(*,4000)IHNMR
       IF(IHNMR .EQ. 1) CALL HNMRIN(IIN,IOUT,IOK,HNMR)
       IF(IOK .EQ. -1) GOTO 7
       WRITE (*,6020)
       READ (*,4000)IC13
 6020 FORMAT(' ARE CARBON 13 SPECTRA AVAILABLE? (01=YES 0=NO)')
       IF(IC13 .EQ. 1) CALL CNMRIN(IIN,IOUT,IOK,CNMR)
       IF(IOK .EQ. -1)GOTO 7
C
C**********************************************************************
C
C     DETERMINE WHICH CALCULATION MODE TO USE
C           MODE 0=NO HNMR OR CNMR DATA
C           MODE 1=NO HNMR DATA, CNMR AVAILABLE
C           MODE 2=HNMR AVAILABLE, NO CNMR
C           MODE 3=HNMR AND CNMR AVAILABLE
       MODE=IHNMR*2+IC13
C**********************************************************************
       DO 20 J=1,4
   20 F(J)=EA(2)*HNMR(J)
       DO 25 J=5,10
   25 F(J)=EA(1)/12.*CNMR(J-4)
       IF(MODE .NE. 0) GOTO 30
       WRITE(*,6030)
 6030 FORMAT(' TERMINATE SAMPLE-INSUFFICIENT DATA')
       GOTO 7
   30 CONTINUE
       CO=EA(3)/16.
       CN=EA(4)/14.
       CS=EA(5)/32.
C
C     S IN DIBENZOTHIOPHENE
C     N IN CARBAZOLE
C     O IN PHENOL
       F(1)=F(1)-8.*CN-8.*CS
       F(6)=F(6)-12.*CN-12.*CS

       IF (MODE .EQ. 1)F(1)=EA(2)
       IF(MODE .EQ. 1)CALL CONC1(IIN,IOUT,MODE,F,MID)
       IF(MODE .EQ. 2)F(5)=EA(1)/12.
       IF(MODE .EQ. 2 .OR. MODE .EQ. 3)CALL CONC(IIN,IOUT,MODE,F,MID)
       WRITE(*,6040)MID,CO,CN,CS
 6040 FORMAT(' HYDROCARBON GROUP CONCENTRATIONS:',/,
      2 '     BENZENE=',E9.3,/,'     NAPHTHALENE=',E9.3,/,
      2 '     ALPHA METHYL=',E9.3,/,'     HYDROAROMATIC=',E9.3,/,
      2 '     BETA CH2=',E9.3,/,'     BETA CH=',E9.3,/,
      2 '     BETA CH3=',E9.3,/,'     ALPHA CH2=',E9.3,/,
      2 '     ALPHA CH=',E9.3,/,'     PHENOL=',E9.3,/,
      2 '     CARBAZOLE=',E9.3,/,'     DIBENZOTHIOPHENE=',E9.3)
       STOP
       END
```

Table A2-6 (continued)

```
C
      SUBROUTINE SARAFG(IIN,IOUT)
      REAL SARA(10),EA(5),F(10),CNMR(6),HNMR(4),G(9),HA(8),TOTHA(8)
      REAL TOTHC(9)
C     SARA DATA AVAILABLE - INPUT YIELDS AND VERIFY SELECTION
C     OF HETEROATOM ASSIGNMENTS
C
C     NOMENCLATURE
C
C     SARA(10)  CONTAINS THE WT% YIELDS OF EACH OF THE 10 SARA FRACTIONS
C     EA(5)     CONTAINS THE ELEMENTAL COMPOSITION DATA (C,H,N,O,S)
C               FOR THE SARA FRACTION CURRENTLY BEING CONSIDERED
C     HA(8)     HETEROATOM CONCENTRATIONS FOR THE CURRENT SARA
C               FRACTION
C     TOTHA(8)  TOTAL HETEROATOM CONCENTRATION IN THE WHOLE OIL
C     F(10)     CONCENTRATIONS OF ATOMIC GROUPS: 1-4 ARE THE VARIOUS
C               H ATOM CONCS., 5-10 ARE THE VARIOUS C ATOM CONCS.
C     CNMR(6)   FRACTIONS OF C IN THE 6 BONDING ENVIRONMENTS
C     HNMR(4)   FRACTIONS OF H IN THE 4 BONDING ENVIRONMENTS
C     G(9)      GROUP CONCENTRATIONS FOR HYDROCARBON GROUPS
C
    3 WRITE (*, 6040)
 6040 FORMAT(' THE FUNCTIONAL GROUP DETERMINATION ALGORITHM USES THE'
     2 ,/,' YIELDS FROM 10 SARA FRACTIONS TO HELP ESTIMATE FUNCTIONAL'
     2 ,/,' GROUP CONCENTRATIONS. THESE FRACTIONS ARE (1) NEUTRAL OILS'
     2 ,/,' (2) ASPHALTENES, (3) NEUTRAL RESINS, (4) VERY WEAK BASES,'
     2 ,',' (5) WEAK BASES, (6) STRONG BASES, (7) VERY WEAK ACIDS,'
     2 ,/,' (8) WEAK ACIDS, (9) STRONG ACIDS, (10) SATURATES.')
      WRITE (*,6050)
      READ (*,4050)SARA(1)
      WRITE (*,6051)
      READ (*,4050)SARA(2)
      WRITE (*,6052)
      READ (*,4050)SARA(3)
      WRITE (*,6053)
      READ (*,4050)SARA(4)
      WRITE (*,6054)
      READ (*,4050)SARA(5)

      WRITE (*,6055)
      READ (*,4050)SARA(6)
      WRITE (*,6056)
      READ (*,4050)SARA(7)
      WRITE (*,6057)
      READ (*,4050)SARA(8)
      WRITE (*,6058)
      READ (*,4050)SARA(9)
      WRITE (*,6059)
      READ (*,4050)SARA(10)
 4050 FORMAT(F6.2)
 6050 FORMAT (' INPUT WT% NEUTRAL OILS (XX.XX)')
 6051 FORMAT (' INPUT WT% ASPHALTENES (XX.XX)')
 6052 FORMAT (' INPUT WT% NEUTRAL RESINS (XX.XX)')
 6053 FORMAT (' INPUT WT% VERY WEAK BASES (XX.XX)')
 6054 FORMAT (' INPUT WT% WEAK BASES (XX.XX)')
 6055 FORMAT (' INPUT WT% STRONG BASES (XX.XX)')
 6056 FORMAT (' INPUT WT% VERY WEAK ACIDS (XX.XX)')
 6057 FORMAT (' INPUT WT% WEAK ACIDS (XX.XX)')
 6058 FORMAT (' INPUT WT% STRONG ACIDS (XX.XX)')
 6059 FORMAT (' INPUT WT% SATURATES (XX.XX)')
```

Table A2-6 (continued)

```
C     WRITE THE RESULTS AND PROMPT USER FOR NMR AND ELEMENTAL ANALYSIS
      SUM=0.
      DO 10 I=1,10
   10 SUM=SUM+SARA(I)
      WRITE(*,6060)SARA,SUM
 6060 FORMAT(' THE FOLLOWING VALUES HAVE BEEN ENTERED',/,
     2 ' NEUTRAL OILS=',F6.2,/,
     2 ' ASPHALTENES=',F6.2,/,
     2 ' NEUTRAL RESINS=',F6.2,/,
     2 ' VERY WEAK BASES=',F6.2,/,' WEAK BASES=',F6.2,/,
     2 ' STRONG BASES=',F6.2,/,
     2 ' VERY WEAK ACIDS=',F6.2,/,' WEAK ACIDS=',F6.2,/,
     2 ' STRONG ACIDS=',F6.2,/,' SATURATES=',F6.2,/, ' SUM=',F6.2,
     2 ' OK? (01=YES, 0=NO, -1=EXIT)')
      READ(*,4001)IOK
 4001 FORMAT(I2)
      IF(IOK .EQ. 0)GOTO 3
      IF(IOK .EQ. -1)RETURN
C
C     THIS IS THE MAIN BODY OF THIS SUBROUTINE.
C     THE 10 FRACTIONS ARE TREATED SEQUENTIALLY.
C     FOR EACH FRACTION, THE ANALYTICAL DATA IS INPUT INTERACTIVELY,
C     THEN THE CONCENTRATIONS FOR THAT FRACTION ARE CALCULATED
C     AND PRINTED OUT.
      WRITE(*,6061)
 6061 FORMAT(' FOR EACH NON-ZERO SARA YIELD, ENTER THE REQUESTED',
     2 ' ANALYTICAL DATA')
      DO 30 I=1,10
      IF(SARA(I) .EQ. 0.) GOTO 30
      IF(I .EQ. 1) WRITE(*,6110)
      IF(I .EQ. 2) WRITE(*,6120)

      IF(I .EQ. 3) WRITE(*,6130)
      IF(I .EQ. 4) WRITE(*,6140)
      IF(I .EQ. 5) WRITE(*,6150)
      IF(I .EQ. 6) WRITE(*,6160)
      IF(I .EQ. 7) WRITE(*,6170)
      IF(I .EQ. 8) WRITE(*,6180)
      IF(I .EQ. 9) WRITE(*,6190)
      IF(I .EQ. 10) WRITE(*,6200)
 6110 FORMAT(' ANALYTICAL DATA FOR NEUTRAL OILS')
 6120 FORMAT(' ANALYTICAL DATA FOR ASPHALTENES')
 6130 FORMAT(' ANALYTICAL DATA FOR NEUTRAL RESINS')
 6140 FORMAT(' ANALYTICAL DATA FOR VERY WEAK BASES')
 6150 FORMAT(' ANALYTICAL DATA FOR WEAK BASES')
 6160 FORMAT(' ANALYTICAL DATA FOR STRONG BASES')
 6170 FORMAT(' ANALYTICAL DATA FOR VERY WEAK ACIDS')
 6180 FORMAT(' ANALYTICAL DATA FOR WEAK ACIDS')
 6190 FORMAT(' ANALYTICAL DATA FOR STRONG ACIDS')
 6200 FORMAT(' ANALYTICAL DATA FOR SATURATES')

      CALL EAIN(IIN,IOUT,IOK,EA)
      IF(IOK .EQ. -1) RETURN
      ICNMR=0
      IHNMR=0
      DO 35 J=1,4
   35 HNMR(J)=0.
      DO 36 J=1,6
   36 CNMR(J)=0.
      WRITE(*,6210)
 6210 FORMAT(' ARE HNMR DATA AVAILABLE FOR THIS FRACTION? (01=YES',
     2 ' 0=NO)')
      READ(*,4001)IHNMR
      IF(IHNMR .EQ. 1) CALL HNMRIN(IIN,IOUT,IOK,HNMR)
      IF(IOK .EQ. -1)RETURN
```

Table A2-6 (continued)

```
      WRITE(*,6220)
 6220 FORMAT(' ARE CNMR DATA AVAILABLE FOR THIS FRACTION?',
     2 ' (01=YES, 0=NO)')
      READ(*,4001)ICNMR
      IF(ICNMR .EQ. 1) CALL CNMRIN(IIN,IOUT,IOK,CNMR)
      IF(IOK .EQ. -1)RETURN
C*******************************************************************
C
C     DETERMINE WHICH CALCULATION MODE TO USE
C          MODE 0=NO HNMR OR OR CNMR DATA
C          MODE 1=NO HNMR DATA, CNMR AVAILABLE
C          MODE 2=HNMR AVAILABLE, NO CNMR
C          MODE 3=HNMR AND CNMR AVAILABLE
      MODE=IHNMR*2+ICNMR
C*******************************************************************
C
C
      IF (MODE .EQ. 0) WRITE(*,6230)
 6230 FORMAT(' IF ONLY ELEMENTAL ANALYSIS DATA ARE AVAILABLE,',/,
     2 ' ONLY HETEROATOM CONCENTRATIONS CAN BE CALCULATED')


C
C     COMPUTE HETEROTOM FUNCTIONALITY CONCENTRATIONS FOR EACH FRACTION
C     AND ADJUST ATOMIC CONCENTRATIONS ACCORDINGLY
C
      CC=EA(1)/12.
      CO=EA(3)/16.
      CN=EA(4)/14.
      CS=EA(5)/32.
      DO 98 J=1,8
   98 HA(J)=0.
      DO 99  J=1,4
   99 F(J)=EA(2)*HNMR(J)
      IF(MODE .EQ. 2)CNMR(1)=1.
      IF(MODE .EQ. 1)F(1)=EA(2)
      DO 100 J=1,6
  100 F(J+4)=CC*CNMR(J)
C     DETERMINE IF SOLUTION IS FEASIBLE BY CALLING LINPR SUBROUTINE
C
C     HYDROCARBON GROUP NUMBERS          HETEROATOM GROUP NUMBERS
C       (G)  1=BENZENE                (HA) 1=DIBENZOTHIOPHENE
C            2=NAPHTHALENE                 2=CARBAZOLE
C            3=ALPHA METHYL                3=DIARYL ETHER
C            4=HYDROAROMATIC               4=PHENOL
C            5=BETA CH2                    5=KETONE
C            6=BETA CH                     6=CARBOXYLIC ACID
C            7=BETA CH3                    7=QUINOLINE
C            8=ALPHA CH2                   8=AMINE
C            9=ALPHA CH
C
C     ATOMIC CONCENTRATIONS
C            1=AROMATIC HYDROGEN
C            2=ALPHA HYDROGEN
C            3=BETA HYDROGEN
C            4=GAMMA HYDROGEN
C            5=CARBOXYL CARBON
C            6=AROMATIC CARBON
C            7=CH CARBON
C            8=CH2 CARBON
C            9=ALPHA CH3 CARBON
C            10=NON-ALPHA CH3 CARBON
C
C
```

224

Table A2-6 (continued)

```
        IF(I .NE. 1) GOTO 101
C       NEUTRAL OIL HETEROATOM ASSIGNMENTS:
C               O IN DIAYRL ETHERS
C               N IN CARBAZOLE
C               S IN DIBENZOTHIOPHENE
        F(1)=F(1)-8.*CS-8.*CN+2.*CO
        F(3)=F(3)-CN
        F(6)=F(6)-12.*CN-12.*CS
        HA(1)=CS
        HA(2)=CN

        HA(3)=CO
        IF (MODE .EQ. 1)CALL CONC1(IIN,IOUT,F,G)
        IF (MODE .EQ. 2 .OR. MODE .EQ. 3)CALL CONC(IIN,IOUT,MODE,F,G)
        GOTO 29
   101 IF(I .NE. 2)GOTO 102
C       ASPHALTENE HETEROATOM ASSIGNMENTS
C       O IN PHENOLS (NOT ALCOHOLS)
C       N IN CARBAZOLE
C       S IN DIBENZOTHIOPHENE
        F(1)-F(1)-8.*CS-8.*CN
        F(2)=F(2)-CN
        F(6)=F(6)-12.*CN-12.*CS
        HA(1)=CS
        HA(2)=CN
        HA(4)=CO
        IF (MODE .EQ. 2 .OR. MODE .EQ. 3)CALL CONC(IIN,IOUT,MODE,F,G)
        IF (MODE .EQ. 1)CALL CONC1(IIN,IOUT,F,G)
        GOTO 29
   102 IF(I .NE. 3)GOTO 103
C       NEUTRAL RESIN HETEROATOM ASSIGNMENTS
C               O IN KETONES AND CARBOXYLIC ACIDS (EQUAL MOLAR CONCS.)
C               N IN CARBAZOLE
C               S IN DIBENZOTHIOPHENE
        F(1)=F(1)-8.*CS-8.*CN
        F(2)=F(2)-CN
        F(5)=F(5)-CO/3.
        F(6)=F(6)-12.*CN-12.*CS
        HA(1)-CS
        HA(2)=CN
        HA(5)=CO/3.
        HA(6)=CO/3.
        IF (MODE .EQ. 2 .OR. MODE .EQ. 3)CALL CONC(IIN,IOUT,MODE,F,G)
        IF (MODE .EQ. 1)CALL CONC1(IIN,IOUT,F,G)
        GOTO 29
   103 IF(I .NE. 4)GOTO 104
C       VERY WEAK BASE HETEROATOM ASSIGNMENTS
C               O IN PHENOLICS
C               N IN CARBAZOLES
C               S IN DIBENZOTHIOPHENES
        F(1)=F(1)-8.*CS-8.*CN
        F(3)=F(3)-CN
        F(6)=F(6)-12.*CN-12.*CS
        HA(1)=CS
        HA(2)=CN
        HA(4)=CO
        IF (MODE .EQ. 2 .OR. MODE .EQ. 3)CALL CONC(IIN,IOUT,MODE,F,G)
        IF (MODE .EQ. 1)CALL CONC1(IIN,IOUT,F,G)
   104 IF(I .NE. 5)GOTO 105
C       WEAK BASE HETEROATOM ASSIGNMENTS
C               O IN PHENOLICS
C               N IN PYRIDINES
C               S IN DIBENZOTHIOPHENES
```

Table A2-6 (continued)

```
      F(1)=F(1)-8.*CS-CN
      F(6)=F(6)-12.*CS-3.*CN
      HA(1)=CS
      HA(7)=CN
      HA(4)=CO
      IF (MODE .EQ. 2 .OR. MODE .EQ. 3)CALL CONC(IIN,IOUT,MODE,F,G)
      IF (MODE .EQ. 1)CALL CONC1(IIN,IOUT,F,G)
      GOTO 29
  105 IF(I .NE. 6)GOTO 106
C     STRONG BASE HETEROATOM ASSIGNMENTS
C         O IN PHENOLICS
C         N IN AMINES AND PYRIDINES (EQUAL CONCENTRATIONS ASSUMED)
C         S IN DIBENZOTHIOPHENES
      F(1)=F(1)+.5*CN-CN-8.*CS
      F(6)=F(6)-12.*CS-1.5*CN
      F(3)=F(3)-CN
      HA(1)=CS
      HA(4)=CO
      HA(7)=CN/2.
      HA(8)=CN/2.
      IF (MODE .EQ. 2 .OR. MODE .EQ. 3)CALL CONC(IIN,IOUT,MODE,F,G)
      IF (MODE .EQ. 0)CALL CONC1(IIN,IOUT,F,G)
  106 IF(I .NE. 2)GOTO 107
C     VERY WEAK ACID HETEROATOM ASSIGNMENTS
C         O IN PHENOLS (NOT ALCOHOLS)
C         N IN CARBAZOLE
C         S IN DIBENZOTHIOPHENE
      F(1)=F(1)-8.*CS-8.*CN
      F(2)=F(2)-CN
      F(6)=F(6)-12.*CN-12.*CS
      HA(1)=CS
      HA(2)=CN
      HA(4)=CO
      IF (MODE .EQ. 2 .OR. MODE .EQ. 3)CALL CONC(IIN,IOUT,MODE,F,G)
      IF (MODE .EQ. 1)CALL CONC1(IIN,IOUT,F,G)
      GOTO 29
  107 IF(I .NE. 8)GOTO 108
C     WEAK ACID HETEROATOM ASSIGNMENTS
C         O IN PHENOLS AND CARBOXYLIC ACIDS (EQUAL MOLAR CONCS.)
C         N IN CARBAZOLE
C         S IN DIBENZOTHIOPHENE
      F(1)=F(1)-8.*CS-8.*CN
      F(2)=F(2)-CN
      F(5)=F(5)-CO/3.
      F(6)=F(6)-12.*CN-12.*CS
      HA(1)=CS
      HA(2)=CN
      HA(4)=CO/3.
      HA(6)=CO/3.
      IF (MODE .EQ. 2 .OR. MODE .EQ. 3)CALL CONC(IIN,IOUT,MODE,F,G)
      IF (MODE .EQ. 1)CALL CONC1(IIN,IOUT,F,G)
      GOTO 29

  108 IF(I .NE. 9)GOTO 109
C     STRONG ACID HETEROATOM ASSIGNMENTS
C         O CARBOXYLIC ACIDS
C         N IN CARBAZOLE
C         S IN DIBENZOTHIOPHENE
      F(1)=F(1)-8.*CS-8.*CN
      F(2)=F(2)-CN
      F(5)=F(5)-CO/2.
      F(6)=F(6)-12.*CN-12.*CS
      HA(1)=CS
      HA(2)=CN
      HA(6)=CO/2.
```

Table A2-6 (continued)

```
        IF (MODE .EQ. 2 .OR. MODE .EQ. 3)CALL CONC(IIN,IOUT,MODE,F,G)
        IF (MODE .EQ. 1)CALL CONC1(IIN,IOUT,F,G)
        GOTO 29
  109 CONTINUE
C       SATURATED HYDROCARBONS
   29 CONTINUE
        DO 210 J=1,8
  210 TOTHA(J)=TOTHA(J)+SARA(I)*HA(J)/100.
        DO 220 J=1,9
  220 TOTHC(J)=TOTHC(J)+SARA(I)*G(J)/100.
        IF(MODE .EQ. 9) WRITE (*,6305)HA
 6305 FORMAT(' HETEROATOMIC GROUP CONCENTRATIONS',/,
      2 '        DIBENZOTHIOPHENE=',E9.3,/,
      2 '        CARBAZOLE=',E9.3,/,'    DIARYL ETHER=',E9.3,/,
      2 '        PHENOL=',E9.3,/,'       KETONE=',E9.3,/,
      2 '        CARBOXYLIC ACID=',E9.3,/,'    QUINOLINE=',E9.3,/,
      2 '        AMINE=',E9.3)
        IF (MODE .EQ. 2 .OR. MODE .EQ. 3) WRITE(*,6310)HA,G
 6310 FORMAT(' HETEROATOMIC GROUP CONCENTRATIONS',/,
      2 '        DIBENZOTHIOPHENE=',E9.3,/,
      2 '        CARBAZOLE=',E9.3,/,'    DIARYL ETHER=',E9.3,/,
      2 '        PHENOL=',E9.3,/,'     KETONE=',E9.3,/,
      2 '        CARBOXYLIC ACID=',E9.3,/,'    QUINOLINE=',E9.3,/,
      2 '        AMINE=',E9.3,/,' HYDROCARBON GROUP CONCENTRATIONS:',/,
      2 '        BENZENE=',E9.3,/,'     NAPHTHALENE=',E9.3,/,
      2 '        ALPHA METHYL=',E9.3,/,'     HYDROAROMATIC=',E9.3,/,
      2 '        BETA CH2=',E9.3,/,'     BETA CH=',E9.3,/,
      2 '        BETA CH3=',E9.3,/,'     ALPHA CH2=',E9.3,/,
      2 '        ALPHA CH=',E9.3)
   30 CONTINUE
        WRITE(*,6385)TOTHA
 6385 FORMAT(' WHOLE OIL HETEROATOMIC GROUP CONCENTRATIONS',/,
      2 '        DIBENZOTHIOPHENE=',E9.3,/,
      2 '        CARBAZOLE=',E9.3,/,'    DIARYL ETHER=',E9.3,/,
      2 '        PHENOL=',E9.3,/,'     KETONE=',E9.3,/,
      2 '        CARBOXYLIC ACID=',E9.3,/,'    QUINOLINE=',E9.3,/,
      2 '        AMINE=',E9.3)
        IF (MODE .NE. 0) GOTO 31
C
C       TO DETERMINE HYDROCARBON CONCENTRATIONS, DATA ON THE WHOLE
C       OIL WILL BE USED
C
        WRITE(*,6015)
 6015 FORMAT(' INPUT DATA ON WHOLE OIL')
        CALL EAIN(IIN,IOUT,IOK,EA)
        IF(IOK .EQ. -1) RETURN
        WRITE(*,6017)
 6017 FORMAT(' ARE HNMR SPECTRA AVAILABLE? (01=YES, 0=NO)')
        READ(*,4001)IHNMR
        IF(IHNMR .EQ. 1) CALL HNMRIN(IIN,IOUT,IOK,HNMR)
        IF(IOK .EQ. -1) RETURN
        WRITE (*,6020)
        READ (*,4001)IC13
 6020 FORMAT(' ARE CARBON 13 SPECTRA AVAILABLE? (01=YES 0=NO)')
        IF(IC13 .EQ. 1) CALL CNMRIN(IIN,IOUT,IOK,CNMR)
        IF(IOK .EQ. -1)RETURN
```

Table A2-6 (continued)

```
C
C***********************************************************************
C
C     DETERMINE WHICH CALCULATION MODE TO USE
C             MODE 0=NO HNMR OR CNMR DATA
C             MODE 1=NO HNMR DATA, CNMR AVAILABLE
C             MODE 2=HNMR AVAILABLE, NO CNMR
C             MODE 3=HNMR AND CNMR AVAILABLE
      MODE=IHNMR*2+IC13
C***********************************************************************
      WRITE(*,9971)MODE,IC13,IHNMR
 9971 FORMAT(1H1,'MODE= 'I2,'IC13 ',I2,'IHNMR ',I2)
      DO 200 J=1,4
  200 F(J)=EA(2)*HNMR(J)
      DO 250 J=5,10
  250 F(J)=EA(1)/12.*CNMR(J-4)
      IF(MODE .NE. 0) GOTO 230
      WRITE(*,6400)
 6400 FORMAT(' TERMINATE SAMPLE-INSUFFICIENT DATA')
      RETURN
  230 CONTINUE
      IF (MODE .EQ. 1)F(1)-EA(2)
      IF(MODE .EQ. 1)CALL CONC1(IIN,IOUT,F,G)
      IF(MODE .EQ. 2 .OR. MODE .EQ. 3)CALL CONC(IIN,IOUT,MODE,F,G)
      WRITE(*,6410)G
 6410 FORMAT(' HYDROCARBON GROUP CONCENTRATIONS:',/,
     2 '      BENZENE=',E9.3,/,'      NAPHTHALENE=',E9.3,/,
     2 ' ALPHA METHYL=',E9.3,/,      HYDROAROMATIC=',E9.3,/,
     2 '     BETA CH2=',E9.3,/,    BETA CH=',E9.3,/,
     2 '     BETA CH3=',E9.3,/,'    ALPHA CH2=',E9.3,/,
     2 '     ALPHA CH=',E9.3)
      RETURN
   31 CONTINUE
      WRITE(*,6410)TOTHC
      RETURN
      END

C
      SUBROUTINE HNMRIN(IIN,IOUT,IOK,HNMR)
C     THIS SUBROUTINE COLLECTS PROTON NMR DATA INTERACTIVELY
      REAL HNMR(4)
    5 WRITE(*,6000)
 6000 FORMAT(' THE PROTON NMR SPECTRUM SHOULD BE DIVIDED INTO THE',/,
     2 ' FOLLOWING 4 BANDS:',/,' GAMMA (0.5-1.0 PPM)',/,
     2 ' BETA (1.0-2.2 PPM)',/,' ALPHA (2.2-5.0 PPM)',/,
     2 ' AROMATIC (5.0-9.0 PPM)')
      WRITE(*,6010)
 6010 FORMAT(' ENTER AROMATIC HYDROGEN FRACTION (0<F<1)')
      READ(*,4000)HNMR(1)
      WRITE(*,6020)
 6020 FORMAT(' ENTER ALPHA HYDROGEN FRACTION (0<F<1)')
      READ(*,4000)HNMR(2)
      WRITE(*,6030)
 6030 FORMAT(' ENTER BETA HYDROGEN FRACTION (0<F<1)')
      READ(*,4000)HNMR(3)
      WRITE(*,6040)
 6040 FORMAT(' ENTER GAMMA HYDROGEN FRACTION (0<F<1)')
      READ(*,4000)HNMR(4)
      SUM=0.
      DO 10 I=1,4
   10 SUM=SUM+HNMR(I)
```

Table A2-6 (continued)

```
      WRITE(*,6050)HNMR,SUM
 6050 FORMAT(' THE FOLLOWING VALUES HAVE BEEN ENTERED:',/,
     2 ' AROMATIC FRACTION=',E12.5,/,' ALPHA FRACTION=',E12.5,/,
     2 ' BETA FRACTION=',E12.5,/,' GAMMA FRACTION=',E12.5,/,
     2 ' SUM=',E12.5,/,' OK? (01=YES, 0=NO, -1=EXIT)')
 4000 FORMAT(F6.2)
      READ(*,4010)IOK
 4010 FORMAT(I2)
      IF(IOK .EQ. 0)GOTO 5
      RETURN
      END
C
      SUBROUTINE EAIN(IIN,IOUT,IOK,EA)
C     THIS SUBROUTINE PERFORMS THE INTERACTIVE INPUT OF ELEMENTAL ANALYSIS
C     DATA.
C
      REAL EA(5)
    5 WRITE (*,6000)
      READ (*,4000) EA(1)
      WRITE (*,6010)
      READ (*,4000) EA(2)
      WRITE (*,6020)
      READ (*,4000) EA(3)
      WRITE (*,6030)
      READ (*,4000) EA(4)
      WRITE (*,6040)
      READ (*,4000) EA(5)
 4000 FORMAT(F6.2)
 6000 FORMAT(' INPUT WT% CARBON (XX.XX)')
 6010 FORMAT(' INPUT WT% HYDROGEN (XX.XX)')
 6020 FORMAT(' INPUT WT% OXYGEN (XX.XX)')
 6030 FORMAT(' INPUT WT% NITROGEN (XX.XX)')
 6040 FORMAT(' INPUT WT% SULFUR (XX.XX)')
      SUM=0.
      DO 10 I=1,5
   10 SUM=SUM+EA(I)
      WRITE(*,6050)EA,SUM
 6050 FORMAT(' THE FOLLOWING VALUES HAVE BEEN ENTERED',/,
     2 ' %C=',F6.2,/,' %H=',F6.2,/,' %0=',F6.2,/,' %N=',F6.2,/,' %S=',
     2 F6.2,/,' SUM=',F6.2,/,' OK? (01=YES, 0=NO -1=EXIT)')
      READ (*,4010)IOK
 4010 FORMAT(I2)
      IF(IOK .EQ. 0) GOTO 5
      RETURN
      END
C
C
      SUBROUTINE CNMRIN(IIN,IOUT,IOK,CNMR)
      REAL CNMR(6)
C     THIS SUBROUTINE COLLECTS CARBON NMR DATA INTERACTIVELY
C
    2 WRITE(*,6000)
 6000 FORMAT(1H1,'THE CARBON NMR SPECTRUM SHOULD BE DIVIDED INTO THE',
     2 /,' FOLLOWING BANDS:',/,
     2 ' 200-160 PPM = CARBOXYL AND CARBONYL',/,
     2 ' 160-60  PPM = AROMATIC CARBON',/,
     2 ' 60-37   PPM = CARBON IN CH',/,
     2 ' 37-22.5 PPM = CH2 CARBON',/,
     2 ' 22.5-20 PPM = METHYL CARBON ALPHA TO AN AROMATIC RING',/,
     2 ' 20-0    PPM = METHYL CARBON NOT ALPHA TO AN AROMATIC RING',/,
     2 ' NOTE: THE CH, CH2 AND CH3 FRACTIONS MAY BE DETERMINED FROM'
     2 /,' INEPT SPECTRA; INPUT INEPT DATA IN THE SAME WAY')
```

Table A2-6 (continued)

```
      WRITE(*,6010)
 6010 FORMAT(' ENTER CARBOXYL/CARBONYL CARBON FRACTION (0<F<1)')
      READ(*,4000)CNMR(1)
      WRITE(*,6020)
 6020 FORMAT(' ENTER AROMATIC CARBON FRACTION (0<F<1)')
      READ(*,4000)CNMR(2)
      WRITE(*,6030)
 6030 FORMAT(' ENTER CH CARBON FRACTION (0<F<1)')
      READ(*,4000)CNMR(3)
      WRITE(*,6040)
 6040 FORMAT(' ENTER CH2 CARBON FRACTION (0<F<1)')
      READ(*,4000)CNMR(4)
      WRITE(*,6050)
 6050 FORMAT(' ENTER ALPHA METHYL CARBON FRACTION (0<F<1)')
      READ(*,4000)CNMR(5)
      WRITE(*,6060)
 6060 FORMAT(' ENTER NON-ALPHA METHYL CARBON FRACTION (1<F<1)')
      READ(*,4000)CNMR(6)
 4000 FORMAT(F6.3)
      SUM=0.
      DO 10 I=1,6
   10 SUM=SUM+CNMR(I)
      WRITE(*,6070)CNMR,SUM
 6070 FORMAT(' THE FOLLOWING VALUES HAVE BEEN ENTERED',/,
     2 ' CARBOXYL/CARBONYL FRACTION',F6.3,/,
     2 ' AROMATIC CARBON FRACTION',F6.3,/,
     2 ' CH CARBON FRACTION',F6.3,/,' CH2 CARBON FRACTION',F6.3,/,
     2 ' ALPHA METHYL CARBON FRACTION',F6.3,/,
     2 ' GAMMA METHYL CARBON FRACTION',F6.3,/,' SUM',F6.3,/,
     2 ' OK? (01=YES, 0=NO -1=EXIT)')
      READ(*,4101)IOK
 4101 FORMAT(I2)
      IF(IOK .EQ. 0)GOTO 2
      RETURN
      END
C     SUBROUTINE CONC(IIN,IOUT,MODE,F,MID)
      REAL MIN,F(10),MID(9),R(4)
     2 ,DUM(9),G(9),Y(5)
      REAL*8 DSEED,RK1,RK2,RK3,RK4
C     DETERMINE IF SOLUTION IS FEASIBLE BY CALLING LINPR SUBROUTINE
C
C
C     GROUP NUMBERS
C           1=BENZENE
C           2=NAPHTHALENE
C           3=ALPHA METHYL
C           4=HYDROAROMATIC
C           5=BETA CH2
C           6=BETA CH
C           7=BETA CH3
C           8=ALPHA CH2
C           9=ALPHA CH
C
C     ATOMIC CONCENTRATIONS
C           1=ALIPHATIC CHAIN BALANCE CONSTRAINT
C           2=AROMATIC HYDROGEN
C           3=ALPHA HYDROGEN
C           4=BETA HYDROGEN
C           5=GAMMA HYDROGEN
C           6=TOTAL CARBON
C
```

Table A2-6 (continued)

```
        RK1=16807.30
        RK2=(2.D0)**31.D0
        CALL SIMPLX (MID,F,MODE,IER)
        IF(IER .EQ. 0) GOTO 30
        IF(IER .EQ. -2)WRITE(*,6302)
 6302 FORMAT(' TERMINAL ERROR IN AN INTERNAL DATA FILE')
        IF(IER .EQ. -1) WRITE(*,6301)IER
 6301 FORMAT(' INFEASIBLE SOLUTION, ERROR RETURN CODE=',I4)
        IOK=-1

        RETURN
C
C    DIRECT SEARCH CALCULATIONS FOR MODES 2 AND 3
C
   30 CONTINUE
        DSEED=123457.D0
        DO 35  I=1,9
        DUM(I)=MID(I)
   35 CONTINUE
        SCARB=F(5)+F(6)+F(7)+F(8)+F(9)+F(10)
        BOUND=SCARB
        MIN=1.0E5
        IVAL=0
        DO 110 IJ=1,50
        DO 100 IK=1,1000
C
C    RANDOM NUMBER GENERATOR
        DO 40 I=1,4
        RK3=DSEED*16807.DO
        RK4=RK2=1.DO
        DSEED=DMOD(RK3,RK4)
   40 R(I)=DSEED/RK2
C
        G(5)=MID(5)+(R(1)-0.5)*BOUND
        G(6)=MID(6)+(R(2)-0.5)*BOUND
        G(8)=MID(8)+(R(3)-0.5)*BOUND
        G(9)=MID(9)+(R(4)-0.5)*BOUND
        Y(1)=F(1)+G(8)+G(9)
        Y(2)=F(2)-2.*G(8)-G(9)
        Y(3)=F(3)=2.*G(5)-G(6)
        Y(4)=F(4)
        Y(5)=SCARB-G(5)-G(6)-G(8)-G(9)
        G(7)=Y(4)/3.
        G(4)=7(3)/4.
        G(3)-(Y(2)-4.*G(4))/3.
        G(2)=(6(1)-Y(5)+G(7)+6.*G(4)+2.*G(3))/(-2)
        G(1)=(Y(5)-G(7)-4.*G(4)-G(3)-10.*G(2))/6.
        DO 70 I=1,9
        IF(G(I) .LT.  0.) GOTO 100
   70 CONTINUE
C     IF((G(6)-G(7)-G(8)/2.+G(9)) .GT. 0.)GOTO 100
        IF (MODE .EQ. 3) FUN=(F(6)-6.*G(I)-10.*G(2))**2+
     2  (F(7)-G(6)-G(9))**2+(F(8)-4.*G(4)-G(5)-G(8))**2+
     2  (F(9)-G(3))**2-(F(10)-G(7))**2
        IF (MODE .EQ. 2)FUN=69.2*G(1)+76.9*G(2)+11.2*G(3)-9.*G(4)
     2 +9.4*G(5)-12.1*G(6)+30.4*G(7)-9.7*G(8)-31.4*G(9)
        IVAL=IVAL+1
        IF (FUN .GT. MIN) GOTO 100
        MIN=FUN
        DO 80 I=1,9
   80 DUM(I)=G(I)
  100 CONTINUE
```

Table A2-6 (continued)

```
      BOUND=.90*BOUND
      DO 106 I=1,9
  106 MID(I)=DUM(I)
  110 CONTINUE
      WRITE(*,6010)IVAL,MIN
 6010 FORMAT(' NUMBER OF FUNCTION EVALUATIONS=',I5,'   MINIMUM=',E12.5)
      RETURN
      END
C
C
C
C
C
      SUBROUTINE CONC1(IIN,IOUT,MODE,F,MID)
      REAL MIN,F(10),MID(9),BOUND(3),R(3)
     2 ,DUM(9),G(9),Y(2)
C     DETERMINE IF SOLUTION IS FEASIBLE BY CALLING LINPR SUBROUTINE
C
C     GROUP NUMBERS
C            1=BENZENE
C            2=NAPHTHALENE
C            3=ALPHA METHYL
C            4=HYDROAROMATIC
C            5=BETA CH2
C            6=BETA CH
C            7=BETA CH3
C            8=ALPHA CH2
C            9=ALPHA CH
C
C     ATOMIC CONCENTRATIONS
C            1=ALIPHATIC CHAIN BALANCE CONSTRAINT
C            2=AROMATIC CARBON
C            3=CARBON IN CH
C            4=CARBON IN CH2
C            5=METHYL CARBON ALPHA TO A RING
C            6=METHYL CARBON NOT ALPHA TO A RING
C            7=TOTAL HYDROGEN
C
      CALL SIMPLX(MID,F,MODE,IER)
      IF(IER .EQ. 0) GOTO 30
      IF(IER .EQ. -2)WRITE(*,6302)
 6302 FORMAT(' TERMINAL ERROR IN AN INTERNAL DATA FILE')
      IF(IER .EQ. -1) WRITE(*,6301)IER
 6301 FORMAT(' INFEASIBLE SOLUTION, ERROR RETURN CODE=',I4)
      RETURN
C
C     DIRECT SEARCH CALCULATIONS FOR MODE 1
C
   30 CONTINUE
      NR=3
      DSEED=123
      DO 35  I=1,9
      DUM(I)=MID(I)
   35 CONTINUE
      BOUND(1)=F(8)/4.
      BOUND(2)=F(8)
      BOUND(3)=F(7)
      MIN=1.0E5
      DO 110 IJ=1,50
      IVAL=0
      DO 100 IK=1,1000
```

232

Table A2-6 (continued)

```
      DO 40 I=1,3
      DSEED=AMOD((DSEED*16807.),(2.**31-1.))
   40 R(I)=DSEED/2.**31
      G(4)=MID(4)+(R(1)-0.5)*BOUND(1)
      G(8)=MID(8)+(R(2)-0.5)*BOUND(2)
      G(9)=MID(9)+(R(3)-0.5)*BOUND(3)
      Y(1)=F(8)-G(8)-4.*G(4)
      Y(2)=F(7)-G(9)
      G(7)=F(10)
      G(5)=F(8)
      G(3)=F(9)
      G(6)=F(7)
      G(2)=(F(6)-F(1)+2.*G(3)+G(6)+2.*G(5)+3.*G(7))/2.
      G(10=(F(1)-8.*G(2)-2.*G(3)-G(6)-2.*G(5)-3.*G(7))
      DO 7 I=1,9
      IF (G(I) .LT. 0.) GOTO 100
    7 CONTINUE
      IF ((G(6)-G(7)-G(8)/2.+G(9)) .GT. 0) GOTO 100
      TEMP1= (Y(1)-G(6)-G(9))**2+(Y(2)-4.*G(4)-G(5)-G(8))**2
      T9=(F(9)-G(3))**2-(F(10)-G(7))**2+(G(6)-G(7)-G(8)/2.+G(9))**2
      FUN=(F(6)-6.*G(1)-10.*G(2))**2 + TEMP1 + T9
      IVAL=IVAL+1
      IF (FUN .GT. MIN) GOTO 100
      MIN=FUN
      DO 80 I=1,9
   80 DUM(I)=G(I)
  100 CONTINUE
      DO 105 I=1,3
  105 BOUND(I)=.80*BOUND(I)
      DO 106 I=1,9
  106 MID(I)=DUM(I)
  110 CONTINUE
      WRITE(*,6010)IVAL,MIN
 6010 FORMAT(' NUMBER OF FUNCTION EVALUATIONS=',I5,'    MINIMUM=',
     2 E12.5)
      RETURN
      END
C
      SUBROUTINE SIMPLX(RMID,F,MODE,IER)
      CHARACTER*5 RNM1,RNM2,CLNM1,CLNM2,BLNK,NEG,POS,SYMB,CDID,RO,
     1        MA,FI,LGE,IBN1(40),IBN2(40),NBN1(40),NBN2(40),
     2        EO,X1,X2,X3,X4,X6,CARB,AROM,ALFA,BETA,GAMA,PHEN,NITR,
     3        SULF,FURA,ALCA,ARCA,X7,X8,X9,CCH,CCH2,ACH3,GCH3,CDBO,HYDR
      REAL    PI(40),NBP(40),XPI(40),RMID(9),F(10)
      REAL    PIVOT,LST,XNBP,FN,CJBAR,X,VALUE,BP(40),RQ(40),B(40,40)

      DATA POS,FIN,FI,RO,MA,NEG /'+','FIN','RH','RO','MA','-'/,
     2     EO,BLNK,ALCA,ARCA /'EO',' ','ALCA','ARCA'/,
     2 CARB,AROM,ALFA,BETA,GAMA /'CARB','AROM','ALFA','BETA','GAMA'/
     2    ,PHEN,FURA,NITR,SULF /'PHEN','FURA','NITR','SULF'/,
     2   X1,X2,X3,X4,X5,X6,X7,X8 /'X1','X2','X3','X4','X5','X6','X7',
     2     'X8'/,
     2 X9,CCH,CCH2,ACH3,GCH3,CDBO,HYDR /'X9','CCH','CCH2','ACH3',
     2     GCH3','CDBO','HYDR'/
C
C     OPEN DATA FILES
C
      IF(MODE .EQ. 2 .OR. MODE .EQ. 3) OPEN(5,FILE='A:MODE2.DAT',
     1 STATUS='OLD',ACCESS='SEQUENTIAL',FORM='FORMATTED')
      IF (MODE .EQ. 1) OPEN(5,FILE='A:MODE1.DAT',
     1 STATUS='OLD',ACCESS='SEQUENTIAL',FORM='FORMATTED')
C     IF (ICASE .EQ. 1) OPEN(5,FILE='A:COAS5.DAT',STATUS='OLD',
C    1 ACCESS='SEQUENTIAL',FORM='FORMATTED')
C
```

Table A2-6 (continued)

```
C       INPUT PROGRAM
C
        IN=5
54323 CONTINUE
        M=0
        N=1
        ISW = 0
        NROWS = 0
        NGE = 0
        NLE = 0
        NEQ = 0
        NEL = 0
        NRHS = 0
        NCOLS = 0
        IER=0
C
C       CLEAR MATRIX TO ZERO
C
        DO 12 I=1,40
        DO 12 J=1,40
   12 B(I,J) = 0.
C
C       READ FIRST CARD - SHOULD BE ROWID
C
        READ (IN,2)CDID
    2 FORMAT (A2)
        IF (CDID .LT. RO) GOTO 3
        IF (CDID .EQ. RO) GOTO 680
        IF (CDID .GT. RO) GOTO 3


    3 CONTINUE
        IER=-2
 3334 RETURN
C
C           READ AND STORE ROWID CARDS
C           GENERATE POS AND NEG SLACKS AS REQUIRED
C
  680 CONTINUE
  681 FORMAT (80X)
  101 FORMAT(A2,8X,A1,1X,A4,A1)
        IF (CDID .EQ. MA) GOTO 504
        IF (CDID .NE. MA) GOTO 103
  504 CONTINUE
        GO TO 104
  103 M=M+1
        NROWS = NROWS+1
        IF (LGE .EQ. POS) GOTO 106
        IF (LGE .NE. POS) GOTO 105
  105 IF (LGE .EQ. NEG) GOTO 108
        IF (LGE .NE. NEG) GOTO 107
  106 IBN1(M)=RNM1
        IBN2(M)=RNM2
        NLE = NLE+1
        BP(M)= 0.
        GO TO 101
  108 IBN1(M) = RNM1
        IBN2(M) = RNM2
        NGE = NGE+1
        BP(M) = -1.0
        B(M,N) = -1.0
  401 NBN1(N) = RNM1
        NBN2(N) = RNM2
        NBP(N) = 0.
        N = N+1
        GO TO 101
```

Table A2-6 (continued)

```
  107 IBN1(M) = RNM1
      IBN2(M) = RNM2
      NEQ = NEQ+1
      BP(M) = -2.0
      GO TO 101
C
C         READ AND STORE FIRST MATRIX ELEMENT
C
  104 READ(IN,195)CDID,CLNM1,CLNM2,RNM1,RNM2,SYMB,VALUE
  195 FORMAT(I2,8X,A4,A1,5X,A4,A1,4X,A1,F10.0)
      GO TO 119
  109 IF (NBN1(N) .EQ. CLNM1) GOTO 600
      IF (NBN1(N) .NE. CLNM1) GOTO 111
  600 IF (NBN2(N) .EQ. CLNM2) GOTO 601
      IF (NBN2(N) .NE. CLNM2) GOTO 111
  601 CONTINUE
  112 DO 113 I=1,M

      IF (IBN1(I) .EQ. RNM1) GOTO 602
      IF (IBN1(I) .NE. RNM1) GOTO 113
  602 CONTINUE
      IF (IBN2(I) .EQ. RNM2) GOTO 603
      IF (IBN2(I) .NE. RNM2) GOTO 113
  113 CONTINUE
      IER=-2
      RETURN
  603 CONTINUE
  114 IF (SYMB .EQ. NEG) GOTO 115
      IF (SYMB .NE. NEG) GOTO 116
  115 B(I,N) = -VALUE
      GO TO 117
  116 B(I,N) =  VALUE
C
C         READ AND STORE MATRIX ELEMENTS
C
  117 READ(IN,195)CDID,CLNM1,CLNM2,RNM1,RNM2,SYMB,VALUE
      NEL = NEL+1
      GO TO 109
  111 N = N+1
      NCOLS = NCOLS+1
      IF (CDID .EQ. FI) GOTO 190
      IF (CDID .NE. FI) GOTO 119
  119 NBN1(N) = CLNM1
      NBN2(N) = CLNM2
  201 IF (SYMB .EQ. NEG) GOTO 203
      IF (SYMB .NE. NEG) GOTO 202
  202 NBP(N) =  VALUE
      GO TO 117
  203 NBP(N) = -VALUE
      GO TO 117
C
C         READ AND STORE RHS ELEMENTS
C
  190 DO 191 I=1,M
  191 RQ(I)=0.
```

Table A2-6 (continued)

```
      GO TO 120
  120 READ(IN,121) CDID,RNM1,RNM2,VALUE
      IF(RNM1 .EQ. HYDR) VALUE=F(1)+F(2)+F(3)+F(4)
      IF(RNM1 .EQ. CARB) VALUE=F(5)+F(6)+F(7)+F(8)+F(9)+F(10)
      IF(RNM1 .EQ. AROM) VALUE=F(1)
      IF(RNM1 .EQ. ALFA) VALUE=F(2)
      IF(RNM1 .EQ. BETA) VALUE=F(3)
      IF(RNM1 .EQ. GAMA) VALUE=F(4)
      IF(RNM1 .EQ. CDBO) VALUE=F(5)
      IF(RNM1 .EQ. ARCA) VALUE=F(6)
      IF(RNM1 .EQ. CCH ) VALUE=F(7)
      IF(RNM1 .EQ. CCH2) VALUE=F(8)
      IF(RNM1 .EQ. ACH3) VALUE=F(9)
      IF(RNM1 .EQ. GCH3) VALUE=F(10)
  121 FORMAT(A2,8X,A4,A1,5X,F10.0)

      IF (CDID .EQ. EO) GOTO 193
      IF (CDID .NE. EO) GOTO 122
  122 DO 124 I=1,M
      IF (IBN1(I) .EQ. RNM1) GOTO 610
      IF (IBN1(I) .NE. RNM1) GOTO 124
  610 IF (IBN2(I) .EQ. RNM2) GOTO 611
      IF (IBN2(I) .NE. RNM2) GOTO 124
  124 CONTINUE
      IER=-2
      RETURN
  611 CONTINUE
  125 RQ(I)=VALUE
      NRHS = NRHS+1
      GO TO 120
  193 N = N-1
C
C     BLANK OUT ARTIFICIAL NAMES
C
      DO 10 I=1,M
      IF(BP(I)+1.0)19,11,10
   11 IBN1(I) = BLNK
      IBN2(I) = BLNK
      GO TO 10
   19 BP(I) = -1.0
      IBN1(I) = BLNK
      IBN2(I) = BLNK
   10 CONTINUE
C
C     ACCUMULATE COUNT OF INFEASIBILITIES
C
      NINF =0
      DO 6000 I=1,M
      IF(BP(I))6001,6000,6000
 6001 NINF = NINF+1
 6000 CONTINUE
C
C     GENERATE INDICATORS FOR MINIMIZATION OF INFEASIBILITY
C
      DO 6101 J=1,N
      XPI(J) =0.
      DO 6101 I=1,M
      IF(BP(I))6102,6101,6101
 6102 XPI(J) = XPI(J)-B(I,J)
 6101 CONTINUE
      DO 6002 I=1,M
 6002 BP(I) = 0.
      IPHASE = 1
```

Table A2-6 (continued)

```
C
C       MAIN ROUTINE
C
C9201 CONTINUE
      IT=0

54325 CONTINUE
C
C       CALCULATE SHADOW PRICES
C
      DO 194 J=1,N
      PI(J) = -NBP(J)
      DO 194 I=1,M
  194 PI(J) = PI(J) + BP(I)*B(I,J)
C
C       SELECT BEST NONBASIS VECTOR
C
 9101 LST = -.0000001
      KCOL = 0
      GO TO (751,552),IPHASE
  751 IF(NINF)54321,54321,552
  552 CONTINUE
      DO 9102 J=1,N
C
C       IGNORE ARTIFICIAL VARIABLES
C
      IF ((NBN1(J) .EQ. BLNK) .AND. (NBN2(J) .EQ. BLNK)) GOTO 9102
      IF ((NBN1(J) .NE. BLNK) .OR. (NBN2(J) .NE. BLNK)) GOTO 651
  651 CONTINUE
      GO TO (6003,6004),IPHASE
 6003 IF(XPI(J)-LST)6005,6006,6006
 6005 KCOL=J
      LST = XPI(J)
      GO TO 9102
 6004 CONTINUE
      IF(PI(J)-LST)9103,9102,9102
 9103 KCOL = J
      LST = PI(J)
 6006 CONTINUE
 9102 CONTINUE
      IF (KCOL)54321,54321,9104
C
C       DETERMINE KEYROW
C
 9104 KROW = 0
      CJBAR = LST
      LST = 1.0E20
      DO 9105 I=1,M
      IF(B(I,KCOL))9105,9105,9106
 9106 RATIO = RQ(I)/B(I,KCOL)
      IF (RATIO-LST)9107,9105,9105
 9107 LST = RATIO
      KROW=I
 9105 CONTINUE
      IF(KROW)9112,9112,9114
C9112 WRITE(6,9113) NBN1(KCOL),NBN2(KCOL)
 9112 IER=-2
C9113 FORMAT(' VARIABLE ',A4,A1,'  UNBOUNDED ')
```

Table A2-6 (continued)

```
          GO TO 54323
     9114 CONTINUE
C
C          TRANSFORM
C
C          DIVIDE BY PIVOT
          PIVOT = B(KROW,KCOL)
          DO 9108 J=1,N
     9108 B(KROW,J) = B(KROW,J)/PIVOT
          RQ(KROW) = RQ(KROW)/PIVOT
          DO 9119 I=1,M
          IF (I-KROW) 9110,9119,9110
     9110 RQ(I) = RQ(I) - RQ(KROW)*B(I,KCOL)
          DO 9109 J=1,N
          IF (J-KCOL) 9111,9109,9111
     9111 B(I,J) = B(I,J) - B(KROW,J)*B(I,KCOL)
     9109 CONTINUE
     9119 CONTINUE
          DO 9300 I=1,M
     9300 B(I,KCOL)   = -B(I,KCOL)/PIVOT
          B(KROW,KCOL) = 1.0/PIVOT
C
C          INTERCHANGE BASIS AND NONBASIS VARIABLES
C
          RNM1 = NBN1(KCOL)
          RNM2 = NBN2(KCOL)
          NBN1(KCOL) = IBN1(KROW)
          NBN2(KCOL) = IBN2(KROW)
          IBN1(KROW) = RNM1
          IBN2(KROW) = RNM2
          LST = NBP(KCOL)
          NBP(KCOL) = BP(KROW)
          BP(KROW) = LST
          IT = IT + 1
          IF ((NBN1(KCOL) .EQ. BLNK) .AND. (NBN2(KCOL) .EQ. BLNK)) GOTO 6200
          IF ((NBN1(KCOL) .NE. BLNK) .OR. (NBN2(KCOL) .NE. BLNK)) GOTO 6201
     6200 NINF = NINF-1
     6201 CONTINUE
C
C          COMPUTE OBJECTIVE FUNCTION
C
          FN = 0.
          DO 9301 I=1,M
     9301 FN = FN + BP(I)*RQ(I)
          GO TO (7000,7001),IPHASE
     7000 SAVE = PI(KCOL)
          DO 7003 J=1,N
          PI(J) = PI(J) - SAVE*B(KROW,J)
          XPI(J) = XPI(J) - CJBAR*B(KROW,J)
     7003 CONTINUE
          PI(KCOL) = -SAVE/PIVOT
          XPI(KCOL) = -CJBAR/PIVOT

          GO TO 7004
     7001 CONTINUE
          DO 9302 J=1,N
     9302 PI(J) = PI(J) - CJBAR*B(KROW,J)
          PI(KCOL) = -CJBAR/PIVOT
     7004 CONTINUE
C          CHECK FOR ESSENTIAL ZERO
          DO 6111 I=1,M
          DO 6111 J=1,N
          X=B(I,J)
```

Table A2-6 (continued)

```
      IF (ABS(X)-.0000001) 6112,6112,6111
 6112 B(I,J) = 0.
 6111 CONTINUE
C
C
C     LOG ITERATION
C
      GO TO 9101
C
C
54321 CONTINUE
      IF (IPHASE-1) 8000,8000,54322
 8000 IPHASE = 2
      IF (NINF) 8003,8003,8004
C8004 WRITE(6,8005)
 8004 IER=-1
C8005 FORMAT('0 SOLUTION INFEASIBLE',/)
      GO TO 54322
 8003 CONTINUE
      GO TO 54325
54322 CONTINUE
C
C     OUTPUT ROUTINE
C
      DO 3033 I=1,M
C
C     COST RANGING
C
      VALUE = 1.0E20
      LST = 1.0E20
      DO 12300 J=1,N
      IF ((NBN1(J) .EQ. BLNK) .AND. (NBN2(J) .EQ. BLNK)) GOTO 12300
      IF ((NBN1(J) .NE. BLNK) .OR. (NBN2(J) .NE. BLNK)) GOTO 12305
12305 CONTINUE
      IF (B(I,J)) 12301,12300,12302
12302 X=PI(J)/B(I,J)
      IF (X-LST) 12303,12300,12300
12303 LST=X
      GO TO 12300
12301 X=-PI(J)/B(I,J)
      IF(X-VALUE)12304,12300,12300
12304 VALUE = X

12300 CONTINUE
      VALUE = BP(I) + VALUE
      IF (IBN1(I) .EQ. X1) RMID(1)=RQ(I)
      IF (IBN1(I) .EQ. X2) RMID(2)=RQ(I)
      IF (IBN1(I) .EQ. X3) RMID(3)=RQ(I)
      IF (IBN1(I) .EQ. X4) RMID(4)=RQ(I)
      IF (IBN1(I) .EQ. X5) RMID(5)=RQ(I)
      IF (IBN1(I) .EQ. X6) RMID(6)=RQ(I)
      IF (IBN1(I) .EQ. X7) RMID(7)=RQ(I)
      IF (IBN1(I) .EQ. X8) RMID(8)=RQ(I)
      IF (IBN1(I) .EQ. X9) RMID(9)=RQ(I)
      LST = BP(I) - LST
 3033 CONTINUE
C
C     WRITE(*,1118)RMID
C1118 FORMAT(' SIMPLEX SOLN',/,5E12.4,/,4E12.4)
C     WRITE(*,1117)F
C1117 FORMAT(' F=',/,5E12.4,/,5E12.4)
      RETURN
      END
```

Table A2-6 (continued)

DATA FILE MODE1.DAT

Column 1

↓

```
ROWID
ROWID       0 HYDR
ROWID       0 CDBO
ROWID       0 ARCA
ROWID       0 CCH
ROWID       0 CCH2
ROWID       0 ACH3
ROWID       0 GCH3
MATRIX
MATRIX      X1          PROFT       1.
MATRIX      X1          ARCA        6.
MATRIX      X1          HYDR        6.
MATRIX      X2          PROFT       1.
MATRIX      X2          ARCA        10.
MATRIX      X2          HYDR        8.
MATRIX      X3          PROFT       1.
MATRIX      X3          ACH3        1.
MATRIX      X3          HYDR        2.
MATRIX      X4          PROFT       1.
MATRIX      X4          CCH2        4.
MATRIX      X4          HYDR        6.
MATRIX      X5          PROFT       1.
MATRIX      X5          CCH2        1.
MATRIX      X5          HYDR        2.
MATRIX      X6          PROFT       1.
MATRIX      X6          CCH         1.
MATRIX      X6          HYDR        1.
MATRIX      X7          PROFT       1.
MATRIX      X7          GCH3        1.
MATRIX      X7          HYDR        3.
MATRIX      X8          PROFT       1.
MATRIX      X8          CCH2        1.
MATRIX      X8          HYDR        1.
MATRIX      X9          PROFT       1.
MATRIX      X9          CCH         1.
RHS
RHS         HYDR        0.00
RHS         CDBO        0.00
RHS         ARCA        0.00
RHS         CCH         0.00
RHS         CCH2        0.00
RHS         ACH3        0.00
RHS         GCH3        0.00
EOF
FIN
```

Table A2-6 (continued)

DATA FILE MODE2.DAT

Column 1

↓

```
ROWID
ROWID     0 CARBN
ROWID     0 AROMH
ROWID     0 ALFAH
ROWID     0 BETAH
ROWID     0 GAMAH
MATRIX
MATRIX    X1        PROFT     1.
MATRIX    X1        CARBN     6.
MATRIX    X1        AROMH     6.
MATRIX    X2        PROFT     1.
MATRIX    X2        CARBN     10.
MATRIX    X2        AROMH     8.
MATRIX    X3        PROFT     1.
MATRIX    X3        CARBN     1.
MATRIX    X3        AROMH     -1.
MATRIX    X3        ALFAH     3.
MATRIX    X4        PROFT     1.
MATRIX    X4        CARBN     4.
MATRIX    X4        AROMH     -2.
MATRIX    X4        ALFAH     4.
MATRIX    X4        BETAH     4.
MATRIX    X5        PROFT     1.
MATRIX    X5        CARBN     1.
MATRIX    X5        BETAH     2.
MATRIX    X6        PROFT     1.
MATRIX    X6        CARBN     1.
MATRIX    X6        BETAH     1.
MATRIX    X7        PROFT     1.
MATRIX    X7        CARBN     1.
MATRIX    X7        GAMAH     3.
MATRIX    X8        PROFT     1.
MATRIX    X8        CARBN     1.
MATRIX    X8        AROMH     -1.
MATRIX    X8        ALFAH     2.
MATRIX    X9        PROFT     1.
MATRIX    X9        CARBN     1.
MATRIX    X9        AROMH     -1.
MATRIX    X9        ALFAH     1.
RHS
RHS       CARBN     0.00
RHS       AROMH     0.00
RHS       ALFAH     0.00
RHS       BETAH     0.00
RHS       GAMAH     0.00
EOF
```

SUBJECT INDEX

242